Springer Theses

Recognizing Outstanding Ph.D. Research

Aims and Scope

The series "Springer Theses" brings together a selection of the very best Ph.D. theses from around the world and across the physical sciences. Nominated and endorsed by two recognized specialists, each published volume has been selected for its scientific excellence and the high impact of its contents for the pertinent field of research. For greater accessibility to non-specialists, the published versions include an extended introduction, as well as a foreword by the student's supervisor explaining the special relevance of the work for the field. As a whole, the series will provide a valuable resource both for newcomers to the research fields described, and for other scientists seeking detailed background information on special questions. Finally, it provides an accredited documentation of the valuable contributions made by today's younger generation of scientists.

Theses may be nominated for publication in this series by heads of department at internationally leading universities or institutes and should fulfill all of the following criteria

- They must be written in good English.
- The topic should fall within the confines of Chemistry, Physics, Earth Sciences, Engineering and related interdisciplinary fields such as Materials, Nanoscience, Chemical Engineering, Complex Systems and Biophysics.
- The work reported in the thesis must represent a significant scientific advance.
- If the thesis includes previously published material, permission to reproduce this must be gained from the respective copyright holder (a maximum 30% of the thesis should be a verbatim reproduction from the author's previous publications).
- They must have been examined and passed during the 12 months prior to nomination.
- Each thesis should include a foreword by the supervisor outlining the significance of its content.
- The theses should have a clearly defined structure including an introduction accessible to new PhD students and scientists not expert in the relevant field.

Indexed by zbMATH.

Daisuke Fujii

Dynamical Properties of Baryon Resonances in the Holographic QCD

Doctoral Thesis accepted by
Osaka University, Osaka, Japan

Springer

Author
Dr. Daisuke Fujii
Japan Atomic Energy Agency
Naka, Ibaraki, Japan

Supervisor
Prof. Atsushi Hosaka
Research Center for Nuclear Physics (RCNP)
Osaka University
Osaka, Japan

Advanced Science Research Center
Japan Atomic Energy Agency
Osaka, Japan

ISSN 2190-5053 ISSN 2190-5061 (electronic)
Springer Theses
ISBN 978-981-97-0754-6 ISBN 978-981-97-0755-3 (eBook)
https://doi.org/10.1007/978-981-97-0755-3

This Springer imprint is published by the registered company Springer Nature Singapore Pte Ltd.
The registered company address is: 152 Beach Road, #21-01/04 Gateway East, Singapore 189721, Singapore

Supervisor's Foreword

In his doctoral study, Daisuke Fujii worked on the structure of baryons; from their static to dynamic properties. Much effort has been devoted to electromagnetic and strong transitions of resonances.

Hadrons are composite and dynamic objects of quarks and gluons. Because of the color SU(3) gauge structure of QCD, the fundamental theory of strong interaction, the direct approach is almost unique by the simulations of discretized fields. This lattice QCD method has been successfully applied to ground state hadrons and their interactions. Yet there are many unsolved problems especially for excited states or resonances, while abundant new states have been found by large scale accelerator experiments. This is the place where models or effective theories become useful. Implementing important features of QCD with inputs from experimental and simulation data, effective theories can be applied to various physics phenomena with predictions.

The theoretical tool that Daisuke uses is a low energy theory for hadrons which is derived from QCD with the aid of the string theory. It is the top-down approach of holographic QCD proposed by Sakai and Sugimoto, resulting in a flavor gauge theory for mesons. Baryons are described by instantons in five dimensional space-time. They are collective excitations of QCD, in sharp contrast with the quark model where baryons are described as single particle excitations of quarks. The collective and single particle natures are the two opposing limits that either one of them dominates a state is determined by QCD dynamics, which should be carefully investigated for each state. Daisuke has chosen the most controversial state, the Roper resonance of the nucleon, and extensively studied its transition amplitudes. A new important has been obtained for electromagnetic transition. A significant improvement is achieved, which could not be possible in the quark model.

In the thesis, Daisuke starts with pedagogic introduction of the Sakai-Sugimoto model. He shows the derivation of the model with a detailed explanation of gauge/gravity holographic correspondence. Then he introduces instantons and their collective quantization for baryons. He also discusses the Skyrme model briefly which is useful to understand baryons as instantons. Effort has been made for the discussions of various form factors. In doing so he points out a subtle problem in the definition of the currents; a consistency between the method faithful to the holographic dictionary and the phenomenology inspired one. Daisuke compared them and found unsolved issues there that should be further investigated for the application of holographic QCD to hadron physics.

To summarize, Daisuke studied the properties of baryon resonances in a holographic model of QCD. He found new results in good agreement with data emphasizing thecollective dynamics of QCD that a naive quark model does not have. He also found difficulty in the computation of currents that should be further worked out. All these results have been achieved by Daisuke's excellent mathematical skills and physics intuitions. This book will provide a cornerstone for further studies of the structure of baryons.

Osaka, Ibaraki, Japan
May 2023

Prof. Atsushi Hosaka

Preface

In this doctoral thesis, various properties of baryon resonances are investigated using the Sakai-Sugimoto model, one of the holographic QCD models. Even though the Roper resonance is one of the most experimentally established baryon resonances, it is difficult to explain its various properties theoretically. We find that the mass formula obtained from the Sakai-Sugimoto model captures the characteristics of the experimental data of Roper resonances well. Therefore, we attempted to calculate other properties of the Roper resonance, especially the electromagnetic transition amplitude and the decay width of the one pion emission. We also tried to do similar analyses for other nucleon resonances ($\Delta(1232), N^*(1535)$).

For this purpose, it is necessary to obtain the baryon wave function and chiral current in the Sakai-Sugimoto model. In the holographic QCD model, baryons appear as D-branes. In particular, in the Sakai-Sugimoto model, this D-brane is identified with an instanton on D8 brane. Therefore, we consider the motion of this instanton in moduli space and quantize it to obtain the wave function of the baryon. This is the conventional method used in the analysis of solitons and is called collective coordinate quantization. After that, we performed calculations and compared them with experimental data using the two currents defined as expectation values by using the GKP-Witten relation and the Noether current of chiral symmetry in the Sakai-Sugimoto model, respectively. On the other hand, some problems exist in the definition of chiral current, therefore we pointed out these problems.

In addition, Roper-like excitations have recently been found in heavy baryons. Therefore, we discuss the extension of the Sakai-Sugimoto model to heavy flavor for the purpose of these analyses.

Osaka, Ibaraki, Japan | Daisuke Fujii
May 2023

Parts of this thesis have been published in the following journal articles

- D. Fujii, and A. Hosaka, "Heavy baryons in holographic QCD with higher dimensional degrees of freedom," Physical Review D 101 (2020) 12, 126008, 10.1103/PhysRevD.101.126008. Copyright 2020 The Authors. This an open access article licensed under the terms of the Creative Commons Attribution 4.0 International license.
- D. Fujii, and A. Hosaka, "Decay properties of Roper resonance in the holographic QCD," Physical Review D 104 (2021) 1, 014022, 10.1103/PhysRevD.104.014022. Copyright 2021 The Authors. This an open access article licensed under the terms of the Creative Commons Attribution 4.0 International license.
- D. Fujii, A. Iwanaka and A. Hosaka, "Electromagnetic transition amplitude for Roper resonance from holographic QCD," Physical Review D 106 (2022) 1, 014010, 10.1103/PhysRevD.106.014010. Copyright 2022 The Authors. This an open access article licensed under the terms of the Creative Commons Attribution 4.0 International license.
- A. Iwanaka, D. Fujii and A. Hosaka, "Decay properties of N(1535) in the holographic QCD," Physical Review D 105 (2022) 11, 114057, 10.1103/PhysRevD.105.114057. Copyright 2022 The Authors. This an open access article licensed under the terms of the Creative Commons Attribution 4.0 International license.

Acknowlegements

First of all, I would like to express my gratitude to my advisor, Prof. Atsushi Hosaka, for guiding my research through enlightening discussions. Without the knowledge from the fundamental to the recent development of hadron physics that I gained from him, I would not have been able to complete my research and write this doctoral dissertation. I would also like to thank Prof. Noriyoshi Ishii for his many helpful suggestions and comments. Some of my work owes to his suggestions. I am also grateful to Prof. Kazuyuki Ogata, who was transferred to Kyushu University in the middle of my doctoral course. The basic knowledge of numerical computation that I learned from him was especially helpful in carrying out my research. I would also like to take this opportunity to thank Prof. Hiroyuki Noumi, Prof. Masayuki Asakawa and Prof. Tatsuma Nishioka for their assistance in sub-reviewing this doctoral thesis.

I would like to thank all RCNP members. In particular, I am very happy to my collaborators, Akihiro Iwanaka (D1 student) and Yuto Kaneko (M2 student) in RCNP Theory group, for many exciting discussions. I would like to appreciate the secretary, Sachiko Karasuyama, for providing us with a comfortable environment for our research life.

Finally, I would like to gratefully thank my family and grandparents, especially Kayoko Fujii and Kazuo Tomita, for their support in every aspect of my life. Without their encouragement, I could not have completed my Ph.D. course.

This work was supported by JST SPRING, Grant Number JPMJSP.

Contents

Chapter 1
Introduction

Abstract Some typical low-lying states of nucleon resonance and the problems associated with them will be discussed. The Sakai-Sugimoto model is then briefly explained and the motivation for using the model to describe nucleon resonance is given. Finally, the structure of this book is presented.

The strong interactions forming the baryons that occupy more than 99% of the visible matter in our universe are described by quantum chromodynamics (QCD). However, due to their non-perturbative nature, the behavior of low-energy QCD is not always well understood. While the properties of the ground state baryons such as masses and magnetic moments of the octet baryons, are thought to be well understood, much part of them are actually governed by the flavor symmetry. The dynamics of low energy QCD is more directly reflected in the excited/resonant states of the baryons [1]. This was the case as we have seen in the developments of the atomic physics where various observations of atomic spectra revealed their origin due to the motion of electrons and their interactions. We expect a similar situation for QCD; from the study of baryon spectra we may be able to extract the information of the constituents, or effective degrees of freedom, that govern the structure of baryons and their interactions. This motivates us to study baryon resonances.

While the first-principles calculations of the ground state have been developed by Lattice QCD, the simulations of resonances is difficult, because resonances are recognized as a continuum (scattering) state. For this reason, investigation by effective models that incorporate the appropriate degrees of freedom have been useful. Among various effective models, the most standard one is the constituent quark model that describes baryon resonances as excitations of quarks that are confined inside the baryons [2–7]. The model reproduces well the properties of the ground state baryons and reasonably well the first resonant states. The model can also predicts further resonant states. However due to its simplicity the predictive power is limited, and for some resonances, the model lead serious discrepancies in comparison with experimental data. Difficulties are in many cases decay properties of resonances, which are the most important dynamical properties, for example, the decay widths and

D. Fujii, *Dynamical Properties of Baryon Resonances in the Holographic QCD*,
Springer Theses, https://doi.org/10.1007/978-981-97-0755-3_1

electromagnetic transition amplitudes [8]. This is because the quark model describe the resonances as a stable particle, although the actual baryon resonances are recognized as poles appearing in the scattering amplitudes of mesons and baryons. The dynamically coupled-channel model (DCC model) is a well-known model that respects the actual resonances [9]. This model explains the properties of resonances very well by using scattering cross-section data as input. However, it is a phenomenological model that requires a large amount of data input.

In this doctoral thesis, the dynamical properties of nucleons are discussed in the Sakai-Sugimoto model, which is the most successful holographic QCD description of low-energy QCD [10, 11]. So far, static properties such as the mass spectra have been investigated in this model by well utilizing the extra-dimensional degrees of freedom, where its success has been shown [12–16]. On the other hand, it is also essential to elucidate the dynamical properties of resonances and their interactions. In this study, as a milestone of the new development of the study of dynamical properties in the holographic model of QCD, the properties of the nucleon resonance are investigated [17, 18]. Moreover, this model describes nucleon as solitons (instantons) [12, 14], whose are resonances expressed as their collective motion excitations, namely stable particles. Here, the extra-dimensional degrees of freedom play an important role. As described below, this extra-dimension includes the degrees of freedom of the meson and its resonances. Then, from the viewpoint of our four-dimensional spacetime, this solitons (instantons), baryon, can be interpreted as a meson-baryon composite system. Considering that the actual nucleon resonance is recognized as the poles of the scattering amplitudes of the meson and baryon, this baryon picture is very interesting. From this point of view, it is worthwhile to investigate various properties of nucleon resonance using this model.

In addition, hadrons with heavy quarks and their resonances have been studied with great interest in recent years [19–24]. In particular, the existence of Roper-like heavy baryons is of intriguing concern [8]. The Roper-like excitations are observed at energies above about 500 MeV from the ground state that shows flavor-independent properties [23]. Therefore, it is desirable to extend the Sakai-Sugimoto model to cases involving heavy flavors [25–35]. This thesis also presents our work on the development of the Sakai-Sugimoto model in this direction [36].

Hadron resonance as a fundamental excitation produced by the QCD vacuum shows various aspects and opens up many interesting research areas. The nature of the resonance states can be not only theoretically understood but also experimentally verified. Therefore, theoretical and experimental investigation of the structure and properties of hadron resonance are expected to reveal the puzzle of low-energy QCD.

1.1 Nucleon Resonances

In the following, we discuss the specific resonances.

1.1.1 The Δ(1232) Resonance

The $\Delta(1232)$ resonance is a resonant state with isospin $3/2$ and spin-parity $3/2^+$ quantum numbers, with a mass of 1232 MeV and a decay width of about 100 MeV [37]. In the picture of the quark model, which describes a nucleon as three quarks, it can be interpreted as an excitation by a magnetic dipole transition that flips the spin-isospin of one quark of the nucleon.

It is the most strongly excited compared to the other nucleon resonances, and decay to a pion and a nucleon with a branching ratio of almost 100%. The quark model, which successfully explains the magnetic moment of the nucleon, was expected to reproduce the $\Delta(1232)$ resonance electromagnetic transition amplitude, but its prediction is much smaller [3, 7, 38] than the amplitude observed in experiments [39]. It has been found that a simple description of the nucleon as a three-body system of quarks is not sufficient to explain this transition amplitude, in which it is important to take into account the meson clouds produced by the strong coupling of the $\Delta(1232)$ resonance to the pion and nucleon [40]. It is now becoming clear that this picture of a meson-baryon composite system is also very important for understanding other nucleon resonances.

1.1.2 Roper Resonance

The Roper resonance is the first excited state of the nucleon, with spin-parity $1/2^+$, a mass of 1440 MeV and a decay width of about $160 \sim 190$ MeV, which is one of the most established resonances since L. D. Roper observed its existence in the 1960s [41]. Nevertheless, there are many unsolved puzzles regarding its structure and properties.

A long-standing controversial puzzle is the issue of the mass of the Roper resonance. Since the establishment of the picture of baryons composed of three constituent quarks, there have been many studies using the non-relativistic quark model. On one hand, the quark model was found to reproduce experimental values of the masses of many nucleon resonances by using harmonic oscillator-type confinement potentials. On the other hand, the mass of the Roper resonance cannot be explained by the quark model picture. Its mass smaller than the negative parity resonance $N^*(1535)$ has attracted great amount of interests because the naive quark model predicts the mass of the Roper resonance much higher than that of the negative parity state.

Turning to the dynamical properties of the Roper resonance, a further problem was unveiled. That is the fact that the quark model cannot reproduce the data obtained from the electromagnetic transitions of the Roper resonance. The electromagnetic excitations of nucleon resonances have long been studied experimentally and theoretically as an important source of information for understanding QCD. The helicity amplitudes extracted from this electro-production data distinguish competing models. In earlier years, the data were insufficiently precise and the amount of helicity amplitude data points was limited. However in recent years, mainly with the advent of the Continuous Electron Beam Accelerator Facility (CEBAF) at the Thomas Jefferson National Accelerator Facility (JLab), a large amount of precise data has been obtained [42–46]. Motivated by this, many theoretical studies [8, 9, 24, 47] have been devoted to the understanding of this process for the Roper resonance, in particular, it has been argued that the quark three-body picture is inappropriate and that it is important to consider the effect of the meson cloud [8, 9].

This is not the only problem with the Roper resonance. An almost vanishing decay width of one pion emission, which is a forbidden process in the limit of zero momentum of the outgoing pion in the non-relativistic quark model, disagrees with the large value of the experimental data. To solve these problems about electromagnetic transition and one pion emission, many theoretical efforts have been devoted. It was pointed out that relativistic effects of the confined quarks at short distance and meson cloud effects at long distance are important to improve the problems [8, 9].

It follows that these two problems about electromagnetic transition and one pion emission of the Roper resonance stem from the properties of the non-relativistic quark model. These transition processes are related to the following matrix element;

$$\left\langle \text{spin} \otimes \text{isospin} \middle| \mathcal{O}_{\text{spin,isospin}} \middle| \text{spin} \otimes \text{isospin} \right\rangle \langle \psi^{N^*} | e^{i\vec{q}\cdot\vec{x}} | \psi^N \rangle \tag{1.1.1}$$

where the $\mathcal{O}_{\text{spin, isospin}}$ is the operator of the spin and isospin, $\psi^{N(N^*)}$ is the wave function of the nucleon (the Roper resonance), $\vec{q}$ is the momentum of a pion or a photon. This transition process is forbidden in the limit of $\vec{q} \to 0$ due to the orthogonality of the wave function. However, the experimental value has a finite value in this limit, which is a contradiction.

In Ref. [47], the prediction for the helicity amplitude of the electromagnetic transition at the real photon point was improved by adding a correction for the effect of internal quark dynamics. Recently, similar results have been obtained for the decay width of one pion emission. The effect of internal quark dynamics, which is important for the solution of this problem, contributes as a relativistic correction term to the matrix elements as follows. Denoting the momentum of the internal quark as $\vec{p}$, we find that the relativistic corrections add the following terms to the matrix elements;

$$\left\langle \text{spin} \otimes \text{isospin} \middle| \mathcal{O}_{\text{spin,isospin}} \middle| \text{spin} \otimes \text{isospin} \right\rangle \langle \psi^{N^*} \middle| \vec{p} e^{i\vec{q}\cdot\vec{x}} \middle| \psi^N \rangle \tag{1.1.2}$$

Due to the internal quark momentum $\vec{p}$, this matrix element is not zero even in the limit of $\vec{q} \to 0$.

The importance of the effect of the meson clouds around the quark is also remarked on for the understanding of the Roper resonance [8, 9]. As discussed below, the baryon picture of the Sakai-Sugimoto model leads us to expect that the effect of meson clouds is incorporated.

1.1.3 Negative Parity Resonance

The negative parity resonance $N^*(1535)$ is the second excited state of the nucleon with a mass slightly larger than the Roper resonance. There has been interest in studying $N^*(1535)$ from several perspectives, as follows. For example, it has been discussed that when chiral symmetry is restored at finite temperature and finite density, a degenerate pair of different parity states are observed, i.e., the existence of chiral partners. The negative parity resonance $N^*(1535)$ is considered to be a reasonable candidate for a nucleon's chiral partner [48–51]. This fact is of great interest because it indicates that $N^*(1535)$ may play an important role in understanding the chiral symmetry of QCD.

On the other hand, this resonance is known to be strongly coupled to ηN, which is an almost exclusive property of this resonance [52–54]. Therefore, the production of $N^*(1535)$ can be identified by observing the η meson. This property also allows the study of $N^*(1535)$ through meson-nucleus bound states. When the η meson-nucleus bound states are observed, the energy and structure of the bound state depend strongly on the interaction between the η meson and the nucleus, which in other words can also be said to reflect the nature of $N^*(1535)$ in the nucleus. Therefore, $N^*(1535)$ is an interesting research issue because it allows us to further understand the nucleon resonance by studying the meson-nucleus bound state. So far, the search for this resonance state has been conducted, which has not yet led to its identification [55, 56], but the above-mentioned interest has led to ongoing intensive research.

While in the quark model nucleon resonances are described as three-quark systems, in several models [57–59], the negative parity state $N^*(1535)$ may be described as composite states of a ground state baryon and a negative parity meson such as $K\Sigma$. The baryon picture of the Sakai-Sugimoto model can be interpreted as a meson-baryon composite system, and the analysis of this doctoral thesis has a very interesting possibility for the understanding of baryon resonances including $N^*(1535)$.

1.2 Sakai-Sugimoto Model

The Sakai-Sugimoto model [10, 11] has been recognized as the holographic QCD that best reconstructs strongly coupled massless QCD in the large N_c limit at low energies. In the holographic QCD, the question is how to realize QCD in the framework of string theory.

After Polchinski pointed out the importance of D-branes [60], the AdS/CFT (anti-de Sitter/conformal field theory) correspondence was conjectured by Maldacena [61]. A non-conformal, supersymmetric four-dimensional pure Yang-Mills (YM) theory and its dual gravity theory were constructed, for example by Witten, by using N_c D4 brane [62, 63]. We see that the open string with two endpoints on this N_c D4-brane corresponds to a $U(N_c)$ gluon field of $N_c \times N_c$ adjoint representations. The field of $U(N_c)$-fundamental representations corresponding to the quark is an open string with one of its two endpoints on this D4-brane. The other endpoint should be placed on the D-brane corresponding to the flavor degrees of freedom. In this way, we can introduce quarks into the pure YM theory. However, due to the addition of the new D-brane, we can no longer use the gravity solution proposed by Witten. In general, it is quite difficult to obtain a gravity solution for such a complex D-brane system. Therefore, it was proposed in Ref. [64], to introduce D-branes corresponding to the degrees of freedom of the flavor as probes (that have a negligible back reaction to the background field). This allows us to incorporate quark degrees of freedom into four-dimensional pure YM theory based on Witten's N_c D4-brane system and its dual gravity theory. Furthermore, an assignment of D-branes that reproduces the chiral symmetry and its spontaneous breaking was proposed by Ref. [10, 11]. This is the Sakai-Sugimoto model used in this thesis.

The gravity theory equivalent to massless QCD can be regarded as an effective action of the flavor gauge field in the five dimensional space (four space-time and one extra dimension), implementing the spontaneous breaking of chiral symmetry [65, 66], allowing many hadron physics predictions to be obtained with simple analytical calculations [10, 11]. For example, from this effective theory of mesons, scalar and vector meson spectra can be obtained, which well reproduce experimental data. It also shows that the model contains a pion, which is an Nambu-Goldstone (NG) boson associated with the spontaneous breaking of the chiral symmetry. This pion is massless, as one would expect from the fact that the Sakai-Sugimoto model is the gravitational theory equivalent of massless QCD. This model also contains the Skyrme model including the Wess-Zumino-Witten (WZW) term and the (axial-) vector meson. Furthermore, the chiral anomaly is reproduced from the Chern Simons (CS) term corresponding to the WZW term in this model. In addition, many other qualitative and quantitative predictions related to hadron physics are possible, such as vector meson dominance, the Kawarabayashi-Suzuki-Riazuddin-Fayyazuddin (KSRF) relation, the pion form factor, the $U(1)_A$ anomaly, and so on. Moreover, surprisingly, there are practically only two parameters in this model. Nevertheless, it has achieved great success in explaining light flavor hadron physics.

1.3 Baryons for Sakai-Sugimoto Model

In the Sakai-Sugimoto model, baryons are analyzed by soliton picture [12, 14]. T. Skyrme constructed the Skyrme model, which describes baryons as solitons of an infinite number of pions, which can explain many properties of baryons [67–71].

On the other hand, the Sakai-Sugimoto model is a hadron effective model of 1+4 dimensional spacetime, which leads the Skyrme model by projecting the model onto four-dimensional spacetime employing Atiyah-Manton's method [72]. Furthermore, the extra dimension of the model naturally accommodates various excited states of mesons. The baryons in the Sakai-Sugimoto model are known to emerge as instantons on the D8 brane [12, 14, 73–75]. The dynamics of baryons is described as the collective motion of instantons/solitons, which is a very different description from the quark model with a single-particle picture. Interestingly, as we will see below, the baryon picture of the Sakai-Sugimoto model is closely related to that of the meson cloud, which has recently been revealed in the study of Roper resonances.

For the purposes of this doctoral thesis, there are several facts that are particularly noteworthy in this model, as described below. First, the model describes baryon resonances as meson-baryon composite systems. Baryon resonances are represented using extra dimensions that play a crucial role in the realization of the description of mesons and their resonant states. When the baryon resonance is viewed from a 4-dimensional space-time perspective, the Lagrangian appears as a meson-baryon composite system. The extra-dimensional degrees of freedom are also used to represent the negative parity excitation, because the meson-baryon (soliton) composite system is critically important for the description of negative-parity states. In the quark model, we describe the negative parity excitation as an orbital excitation of a single quark. This model, which describes baryons as solitons (instantons), is similar to the baryon picture of the Skyrme model, but it is known that it is generally difficult to deal with negative parity excitation in the Skyrme model. One way is to introduce meson fluctuations around the soliton solutions i.e., meson-baryon composite system [57, 58]. Therefore, one of the unique points of this model is that it can describe the negative parity excitation by utilizing the extra dimensional degrees of freedom. Furthermore, in the obtained mass formula, the masses of the Roper resonance and the negative parity state are degenerate [12]. This is a good feature of the hadron resonance mass spectra than the quark model.

In addition to the discussions in the light flavor sector, recent experiments have also discovered Roper-like states in heavy flavors, which has stimulated some theoretical works. Under these situations, studies on the extended the Sakai-Sugimoto model to heavy flavors have been performed in Ref. [31, 36] and others. In particular, in our study Ref. [36], we find that the extra dimension again plays an important role in introducing heavy flavors to the Sakai-Sugimoto model.

Further studies of the static properties of baryons in this model have been conducted by [13, 16]. For this purpose, it is necessary to define the chiral current using this model. There are two ways to define the chiral current in this model. One is to define the current using the GKP-Witten relation from the standpoint of the AdS/CFT correspondence [16], and the other is to obtain the current as the Noether current from the standpoint of the hadron effective model [13]. The former method is a natural definition of current from the viewpoint of AdS/CFT correspondence, but it causes problems in the analysis of baryons because the current is defined as coupling with the external field in the extra-dimensional boundary. The baryons in this model are identified with instanton solutions in a 3+1 dimensional space including one extra dimension, but since instanton solutions usually have spatial SO(4) symmetry, the current becomes zero when the baryons exist. This is because the classical solution is obtained by ignoring the effect of the warp factor. In Ref. [16], they obtained an asymptotic solution at the boundary that incorporates the effect of the warp factor leading to a well-defined current evaluation. They used this current to investigate the static properties of the baryon and found that it roughly captured the experimental data. On the other hand, almost at the same time as Ref. [16], a study of nucleon resonance with the current obtained from the latter definition was carried out and found to roughly reproduce the experimental data. However, since this current has non-uniqueness in its form, they determine this non-uniqueness so that the chiral current of the Skyrme model is derived when reduced to four-dimensional space-time.

Using these currents, we investigate the dynamical properties of baryons, in particular, electromagnetic transition amplitudes and the decay width of one pion emission. They are related to the following matrix elements, using a certain function f for the soliton picture of the baryon.

$$\langle \text{spin} \otimes \text{isospin} \big| \mathcal{O}_{\text{spin,isospin}} \big| \text{spin} \otimes \text{isospin} \rangle \langle \psi_\rho^{N^*} \big| f(\vec{x}, \rho) e^{i\vec{q}\cdot\vec{x}} \big| \psi_\rho^{N} \rangle \quad (1.3.1)$$

where $\psi_\rho^{N^*}$ is the wave function for instanton monopole vibration of size ρ. Unlike (1.1.1), this matrix element has a nontrivial form and thus has the potential to solve problems concerning the dynamical properties of the Roper resonance from a different perspective in comparison to relativistic corrections. In particular, analysis of the properties of baryon resonance by the Sakai-Sugimoto model is essential to verify the validity of the description of baryon resonance and to understand its phenomenological meaning. In this doctoral thesis, we will explain both current definitions and calculate the physical quantities. The results will show that the latter definition is more effective for a comprehensive analysis of baryon resonances.

1.4 Construction of the Doctoral Thesis

The role of each chapter is described below. Chapter 2 reviews the analysis of mesons and baryons in the Sakai-Sugimoto model. In particular, we emphasize that the Sakai-Sugimoto model has a meson cloud picture. In addition, the resonances of baryons and their dynamics are expressed as the collective motion of instantons/solitons. By explaining this point, we will clarify the baryon picture as a meson-baryon complex system. In addition, Roper-like excitations have recently been found in heavy flavor baryons. Therefore, we first review the treatment of flavor SU(3) including s quark and then discuss the extensions that introduce heavy flavor to the Sakai-Sugimoto model based on Chap. 3. Furthermore, we define the chiral current to analyze the dynamical properties of the baryon resonance. Therefore, in Chap. 4, we will first discuss the definition of the chiral current, and then calculate various physical quantities and compare them with experimental data. Finally, we summarize and discuss prospects in Chap. 5.

References

1. Burkert VD, Lee TSH (2004) Electromagnetic meson production in the nucleon resonance region. Int J Mod Phys E 13:1035–1112
2. Isgur N, Karl G (1978) P wave Baryons in the quark model. Phys Rev D 18:4187
3. Isgur N, Karl G (1979) Positive parity excited Baryons in a quark model with hyperfine interactions. Phys Rev D 19:2653. [Erratum: Phys Rev D 23, 817 (1981)]
4. Isgur N, Karl G (1979) Ground state baryons in a quark model with hyperfine interactions. Phys Rev D 20:1191–1194
5. Hey AJG, Kelly RL (1983) Baryon spectroscopy. Phys Rept 96:71
6. Capstick S (1992) Photoproduction amplitudes of P(11) and P(33) baryon resonances in the quark model. Phys Rev D 46:1965–1972
7. Capstick S, Keister BD (1995) Baryon current matrix elements in a light front framework. Phys Rev D 51:3598–3612
8. Burkert VD, Roberts CD (2019) Colloquium: roper resonance: toward a solution to the fifty year puzzle. Rev Mod Phys 91(1):011003
9. Suzuki N, Julia-Diaz B, Kamano H, Lee TSH, Matsuyama A, Sato T (2010) Disentangling the dynamical origin of P-11 nucleon resonances. Phys Rev Lett 104:042302
10. Sakai T, Sugimoto S (2005) Low energy hadron physics in holographic QCD. Prog Theor Phys 113:843–882
11. Sakai T, Sugimoto S (2005) More on a holographic dual of QCD. Prog Theor Phys 114:1083–1118
12. Hata H, Sakai T, Sugimoto S, Yamato S (2007) Baryons from instantons in holographic QCD. Prog Theor Phys 117:1157
13. Hata H, Murata M, Yamato S (2008) Chiral currents and static properties of nucleons in holographic QCD. Phys Rev D 78:086006
14. Nawa K, Suganuma H, Kojo T (2007) Baryons in holographic QCD. Phys Rev D 75:086003
15. Nawa K, Suganuma H, Kojo T (2007) Brane-induced skyrmions: Baryons in holographic QCD. Prog Theor Phys Suppl 168:231–236
16. Hashimoto K, Sakai T, Sugimoto S (2008) Holographic Baryons: static properties and form factors from gauge/string duality. Prog Theor Phys 120:1093–1137
17. Fujii D, Hosaka A (2021) Decay properties of Roper resonance in the holographic QCD. Phys Rev D 104(1):014022
18. Fujii D, Iwanaka A, Hosaka A (2022) Electromagnetic transition amplitude for Roper resonance from holographic QCD. Phys Rev D 106(1):014010
19. Yamaguchi Y, Hosaka A, Takeuchi S, Takizawa M (2020) Heavy hadronic molecules with pion exchange and quark core couplings: a guide for practitioners. J Phys G 47(5):053001
20. Hosaka A, Iijima T, Miyabayashi K, Sakai Y, Yasui S (2016) Exotic hadrons with heavy flavors: X, Y, Z, and related states. PTEP 2016(6):062C01
21. Chen H-X, Chen W, Liu X, Zhu S-L (2016) The hidden-charm pentaquark and tetraquark states. Phys Rept 639:1–121
22. Tanabashi M et al (2018) Review of particle physics. Phys Rev D 98(3):030001
23. Arifi AJ, Nagahiro H, Hosaka A, Tanida K (2020) Roper-like resonances with various flavor contents and their two-pion emission decays. Phys Rev D 101(11):111502
24. Arifi AJ, Suenaga D, Hosaka A (2021) Relativistic corrections to decays of heavy baryons in the quark model. Phys Rev D 103(9):094003
25. Liu Y, Zahed I (2017) Holographic heavy-light chiral effective action. Phys Rev D 95(5):056022
26. Liu Y, Zahed I (2017) Heavy–light mesons in chiral AdS/QCD. Phys Lett B 769:314–321
27. Liu Y, Zahed I (2017) Heavy and strange holographic baryons. Phys Rev D 96(5):056027
28. Liu Y (2017) Heavy baryons and their exotics from instantons in holographic QCD. Phys Rev D 95(11)
29. Liu Y (2017) Holographic heavy-light chiral effective action. Phys Rev D 95(5)
30. Liu Y (2017) Heavy and strange holographic baryons. Phys Rev D 96(5)

31. Liu Y, Zahed I (2017) Heavy baryons and their exotics from instantons in holographic QCD. Phys Rev D 95(11):116012
32. Li S (2017) Holographic heavy baryons in the witten-sakai-sugimoto model with the d0-d4 background. Phys Rev D 96(10)
33. Liu Y, Nowak MA, Zahed I (2021) Holographic charm and bottom pentaquarks. I. Mass spectra with spin effects. Phys Rev D 104(11):114021
34. Liu Y, Nowak MA, Zahed I (2021) Holographic charm and bottom pentaquarks. II. Open and hidden decay widths. Phys Rev D 104(11):114022
35. Liu Y, Mamo KA, Nowak MA, Zahed I (2021) Holographic charm and bottom pentaquarks. III. Excitations through photoproduction of heavy mesons. Phys Rev D 104(11):114023
36. Fujii D, Hosaka A (2020) Heavy baryons in holographic QCD with higher dimensional degrees of freedom. Phys Rev D 101(12):126008
37. Anderson HL, Fermi E, Long EA, Nagle DE (1952) Total cross-sections of positive pions in hydrogen. Phys Rev 85:936
38. Capstick S (1992) Photoproduction and electroproduction of nonstrange baryon resonances in the relativized quark model. Phys Rev D 46:2864–2881
39. Montanet L et al (1994) Review of particle properties. Part Data Group Phys Rev D 50:1173–1823
40. Sato T, Lee T-SH (1996) Meson exchange model for pi N scattering and gamma N–> pi N reaction. Phys Rev C 54:2660–2684
41. David Roper L (1964) Evidence for a P-11 pion-nucleon resonance at 556 MeV. Phys Rev Lett 12:340–342
42. Aznauryan IG et al (2008) Electroexcitation of the Roper resonance for 1.7 < Q**2 < 4.5 -GeV2 in vec-ep —> en pi+. Phys Rev C 78:045209
43. Aznauryan IG et al (2009) Electroexcitation of nucleon resonances from CLAS data on single pion electroproduction. Phys Rev C 80:055203
44. Dugger M et al (2009) pi+ photoproduction on the proton for photon energies from 0.725 to 2.875-GeV. Phys Rev C 79:065206
45. Mokeev VI et al (2012) Experimental Study of the $P_{11}(1440)$ and $D_{13}(1520)$ resonances from CLAS data on $ep \to e'\pi^+\pi^- p'$. Phys Rev C 86:035203
46. Mokeev VI et al (2016) New Results from the Studies of the $N(1440)1/2^+$, $N(1520)3/2^-$, and $\Delta(1620)1/2^-$ Resonances in Exclusive $ep \to e'p'\pi^+\pi^-$ Electroproduction with the CLAS Detector. Phys Rev C 93(2):025206
47. Kubota T, Ohta K (1976) Relativistic corrections to the baryon resonance photoexcitation amplitudes in the quark model. Phys Lett B 65:374–376
48. Detar CE, Kunihiro T (1989) Linear σ model with parity doubling. Phys Rev D 39:2805
49. Nemoto Y, Jido D, Oka M, Hosaka A (1998) Decays of 1/2- baryons in chiral effective theory. Phys Rev D 57:4124–4135
50. Jido D, Oka M, Hosaka A (2001) Chiral symmetry of baryons. Prog Theor Phys 106:873–908
51. Jido D, Nemoto Y, Oka M, Hosaka A (2000) Chiral symmetry for positive and negative parity nucleons. Nucl Phys A 671:471–480
52. Haider Q, Liu LC (1986) Formation of an η mesic nucleus. Phys Lett B 172:257–260
53. Liu LC, Haider Q (1986) Signature for the existence of η mesic nucleus. Phys Rev C 34:1845–1854
54. Chiang HC, Oset E, Liu LC (1991) The width of bound eta in nuclei. Phys Rev C 44:738–746
55. Chrien RE et al (1988) Search for bound states of the η meson in light nuclei. Phys Rev Lett 60:2595–2598
56. Nagahiro H, Jido D, Hirenzaki S (2009) Formation of eta-mesic nuclei by (pi, N) reaction and N*(1535) in medium. Phys Rev C 80:025205
57. Callan CG Jr, Klebanov IR (1985) Bound state approach to strangeness in the skyrme model. Nucl Phys B 262:365–382
58. Callan CG Jr, Hornbostel K, Klebanov IR (1988) Baryon masses in the bound state approach to strangeness in the skyrme model. Phys Lett B 202:269–275

59. Ezoe T, Hosaka A (2020) $\Lambda(1405)$ as a $\overline{K}N$ Feshbach resonance in the Skyrme model. Phys Rev D 102(1):014046
60. Polchinski J (1995) Dirichlet branes and ramond-ramond charges. Phys Rev Lett 75:4724–4727
61. Juan Martin Maldacena (1998) The large N limit of superconformal field theories and supergravity. Adv Theor Math Phys 2:231–252
62. Witten E (1998) Anti-de Sitter space, thermal phase transition, and confinement in gauge theories. Adv Theor Math Phys 2:505–532
63. Aharony O (2002) The NonAdS/nonCFT correspondence, or three different paths to QCD. In: NATO advanced study institute and EC summer school on progress in string, field and particle theory, pp 3–24, Dec 2002
64. Karch A, Katz E (2002) Adding flavor to AdS / CFT. JHEP 06:043
65. Nambu Y, Jona-Lasinio G (1961) Dynamical model of elementary particles based on an analogy with superconductivity. II. Phys Rev 124:246–254
66. Nambu Y, Jona-Lasinio G (1961) Dynamical model of elementary particles based on an analogy with superconductivity. 1. Phys Rev 122:345–358
67. Skyrme THR (1962) A unified field theory of mesons and baryons. Nucl Phys 31:556–569
68. Zahed I, Brown GE (1986) The Skyrme Model Phys Rep 142:1–102
69. Witten E (1979) Baryons in the 1/n expansion. Nucl Phys B 160:57–115
70. Witten E (1983) Global aspects of current algebra. Nucl Phys B 223:422–432
71. Witten E (1983) Current algebra, baryons, and quark confinement. Nucl Phys B 223:433–444
72. Atiyah MF, Manton NS (1989) Skyrmions from instantons. Phys Lett B 222:438–442
73. Witten E (1998) Baryons and branes in anti-de Sitter space. JHEP 07:006
74. Douglas MR (1999) Branes within branes. NATO Sci. Ser. C 520:267–275
75. Suganuma H, Hori K (2020) Topological objects in holographic QCD. Phys Scripta 95(7):074014

Chapter 2
Hadrons in the Sakai-Sugimoto Model

Abstract A brief introduction to the gauge/gravity (string) correspondence is provided, then we overview the theoretical development of this correspondence until it becomes applicable to the analysis of QCD. Using the Sakai-Sugimoto model resulting from these efforts, we give a method for obtaining the mass spectrum of mesons. In addition, we provide a description of baryons in this model. For this purpose, we will explain the collective coordinate quantization of solitons.

Since the gauge/gravity (string) correspondence was conjectured, there have been many attempts to use this correspondence to elucidate the non-perturbative effects of QCD. The Sakai-Sugimoto model has attracted attention as the holographic QCD that most reproduces low-energy QCD phenomena [1, 2].

2.1 The Gauge/Gravity (String) Correspondence

For this section, we briefly describe the gauge/gravity (string) correspondence. In the next section, we will introduce the holographic QCD model that best explains low-energy QCD, the Sakai-Sugimoto model.

First, we give the least required introduction to superstring theory. Superstring theory provides a unified description of gravity and gauge theory. This is because there are naturally two possible types of strings. One is a closed string whose ends are closed and looped, and the other is an open string whose ends have two endpoints. In the low energy region, where the string can be regarded as a point particle, the properties of particles are reflected in differences in oscillations of these strings, indicating that the closed string is a graviton and the open string is a gauge particle. Closed and open strings can interact to transit with each other, and the consistency of the quantization of the strings shows that the space-time of superstring theory can be formulated only in ten dimensions.

Now consider a superstring theory that consists of only closed strings. At low energy, this theory is known to be a supergravity theory, and it is known that there

D. Fujii, *Dynamical Properties of Baryon Resonances in the Holographic QCD*, Springer Theses, https://doi.org/10.1007/978-981-97-0755-3_2

are several vacuum solutions, in other words, several ten-dimensional spacetime structures. The most trivial solution is flat spacetime. Not only that there is a black-p-brane solution, which is a ten-dimensional spacetime that corresponds to a four-dimensional black hole, with a charge and a horizon extending into $p + 1$ dimensions. For example, in the case of $p = 3$, it is represented by the following metric;

$$ds^2 = H(r)^{-1/2}(-dt^2 + d\vec{x}^2) + H(r)^{1/2}\left(dr^2 + r^2 d\Omega_5^2\right)$$
$$H(r) = 1 + \frac{r_0^4}{r^4}, \; r_0 = (4\pi g_s N_c)^{1/4} l_s, \tag{2.1.1}$$

where l_s is string length, g_s coupling constant of string and there exists a horizon at $r = 0$, (derived by Appendix 2.5.2). The constant r_0 is a typical length scale that characterizes the curvature of spacetime which is related to string length l_s. The charge carried by the black-p-brane is called the Ramond-Ramond charge (RR charge), which is coupled to a tensor field (gauge field) called the Ramond-Ramond field (RR field). Hence, the black-p-brane solution is the source of the graviton and RR fields. From here, the most important work towards the gauge/gravity (string) correspondence was done by Polchinski [3]. He showed that a D-p-brane placed on a flat spacetime, i.e., spreading in $p + 1$ dimensions, and an object that is an endpoint of an open string is equally a source of graviton and RR fields. It should be emphasized here that at the same time, an open string that behaves as a gauge field was introduced.

This fact implies the equivalence of a theory with a D-p-brane placed in flat spacetime and a string theory in black-p-brane spacetime. From here, we have one more thing to consider to elevate this conjecture to the equivalence of gauge theories and gravity (string) theories. Now, the former theory consists of a gauge field corresponding to an open string and a closed string. In order to discuss the correspondence with the gauge theory, we need to extract only the degrees of freedom of the gauge theory on the D-p-brane. For this purpose, we consider the following limit (decoupling limit) which decouples the closed strings from the gauge fields;

$$g_{YM}^2 = (2\pi)^{p-2} g_s l_s^{p-3} = \text{fixed}, \; l_s \to 0. \tag{2.1.2}$$

Then, the coupling constant $\kappa = (2\pi)^{7/2} g_s l_s^4/\sqrt{2}$ of the open and closed strings are zero, decoupling them ($g_{YM}^2 = (2\pi)^{p-2} g_s l_s^{p-3}$ is derived in the Appendix 2.5.1 and this formula of κ is explained in the Appendix 2.5.2.). Next, consider keeping the energy scale of the physical quantity related to the open string finite under the decoupling limit. Now, if one of the multiple D-p-branes is separated by δl in parallel, the open string stretched between the separated branes gains a mass of $\delta l / l_s^2$. This mass corresponds to the energy scale that should be kept finite. To keep this mass finite under the decoupling limit, we need to set $\delta l \to 0$. If we replace the overlapping D-p-brane with a black-p-brane, we see that this limit, $\delta l \to 0$ with $\delta l / l_s^2 = \text{fixed}$, corresponds to considering a near horizon, $r \to 0$ with $u = r / l_s^2 = \text{fixed}$ (This limit manipulation allows us to derive a new expression with u for the black 3-brane metric near the horizon in Appendix 2.5.3.). Based on the above observations, Maldacena proposed the equivalence between the gauge theory on the D-p-brane and the gravity theory in the near horizon of the black-p-brane [4]. This is roughly how we reached the gauge/gravity (string) correspondence.

In the actual analysis, technical problems require the following further condition. Instead of considering superstring theory as a spacetime equivalent to D-p-brane, we have considered a supergravity theory in which the string is treated as a point particle. The condition for this treatment to be viable is that the string length must be sufficiently smaller than the considering energy scale. The string length l_s correction in supergravity theory is expressed as

$$\text{(Gravity theory)} + \mathcal{O}(\mathcal{R}l_s^2). \tag{2.1.3}$$

This means $\mathcal{R}l_s^2 \ll 1$. As concretely shown in the description of the Sakai-Sugimoto model, this corresponds to the limit where the 't Hooft coupling is large, i.e., the case of $\lambda \equiv g_{YM}^2 N_c \gg 1$.

Furthermore, in the explanation so far, we have considered the classical theory of gravity, but a quantum correction to the supergravity theory must be suppressed to justify this treatment. It is known that this quantum correction is added as follows;

$$\text{(Classical gravity theory)} + \mathcal{O}(\mathcal{R}^4 G_{10}), \tag{2.1.4}$$

where $G_{10} = (2\pi)^7 g_s^2 l_s^8/(16\pi)$ is the gravitational constant in ten dimensions, and this correction is known to correspond to an expansion of $1/N_c^2$ in terms of gauge theory.

2.2 The Sakai-Sugimoto Model

In order to adapt this gauge/gravity (string) correspondence to the QCD analysis, the following points need to be further resolved

1. Introduce quarks of fundamental representation
2. Break the supersymmetry
3. Break the conformal invariance and introduce Λ_{QCD}

For the items 2 and 3, an idea was proposed by Witten to solve them [25]. Let us consider a $4+1$ dimensional gauge theory on an N_c D4-brane. The extra dimension is S^1 compactified. Assuming the compactification radius as R_{KK}, this theory can be regarded as a $3+1$-dimensional gauge theory to a good approximation in energy regions sufficiently lower than $1/R_{KK} = M_{KK}$. Gauge theories on D4-branes were not initially conformally invariant and in the present case R_{KK} is introduced as the scale in this case. This R_{KK} plays the role of $\Lambda_{\rm QCD}$. In addition, for the S^1 compactification direction, supersymmetry can be broken by imposing periodic boundary conditions on the boson and anti-periodic boundary conditions on the fermion. In this way, we can construct a strongly coupled $SU(N_c)$ large N_c pure Yang-Mills theory at low energy.

Introduce Quarks

To introduce quarks as fundamental representations of $SU(N_c)$, we perform the following procedures. In the setup so far, we only have fields of adjoint representations of $SU(N_c)$, i.e., gluons and their superpartners. In order to approach QCD, it is necessary to incorporate the fundamental representations of $SU(N_c)$ gauge groups into the theory. The reason why only adjoint representations exist is that the endpoints of open strings only lie on D4-branes. To introduce the fundamental representation, we need to consider the situation where only one endpoint is on the D4-brane. Then, the other endpoint should also be on the D-brane, which introduces a flavor D-brane that plays the role as the other endpoint [5]. Quarks and antiquarks are distinguished by the orientation of the open string. This is because the color charge of the endpoints on the D4-brane is reversed when the orientation is altered. At this time, a new open string with both ends on the flavor D-brane also arises. This degree of freedom is decoupled in the near-horizon limit like the closed string. The open string corresponding to a quark is also the fundamental representation of the $U(N_f)$ gauge group because it has endpoints on the flavor D-brane. The flavor symmetry of QCD is a global symmetry, which means that the quark we have just introduced has a different symmetry than ordinary QCD. However, as mentioned above, the adjoint representation that guarantees gauge symmetry on the flavor D-brane is decoupled in the near horizon limit, so this flavor symmetry can be regarded as a global symmetry.

Here, it is necessary to restrict the flavor D-brane to a combination of branes such that the ground state that appears by quantizing the open string stretched between the color D-brane and the color D-brane is stable.

The Gravity Dual and Probe Approximation

It is generally difficult to obtain the gravitational duality of a complex brane system in which there are two types of branes with color and flavor degree of freedom respectively. Therefore, we consider a situation in which a flavor D-brane is put as a probe in the gravity dual spacetime of a color D-brane. This flavor D-brane should deform the background spacetime, but since the contribution is proportional to N_f, we can neglect the effect of the flavor D-brane on the background spacetime in the case of $N_c \gg N_f$. Such a treatment of the brane is called probe approximation [5]. On the other hand, the flavor D-brane is influenced by the background spacetime and its induced metric needs to be considered. Recalling that the color D-brane carries gluons and the flavor D-brane carries quarks, this treatment corresponds to an approximation in which the gluons affect the quarks but the quarks do not affect the gluons, which, in Lattice QCD terms called the quench approximation.

2.2.1 The Brane Construction

Now that we have described how to introduce quarks, we next need to introduce a proper flavor D-brane. At this time, it is necessary to introduce a flavor D-brane that reproduces the chiral symmetry and its breaking, which are important for low-energy phenomena in QCD.

The Sakai-Sugimoto model consists of N_c D4 branes and N_f D8, $\overline{\text{D8}}$ branes in type IIA superstring theory [1, 2]. Table 2.1 shows the brane configurations in the model. The numbers label the direction of each axis of the ten-dimensional spacetime, and the symbols $\circ$ and $\times$ indicate whether the brane is extended in that axis direction or not. Please refer to Fig. 2.1 together.

In this brane configuration, under the decoupling limit, the undesired degrees of freedom are decoupled and low-energy massless QCD is realized. Under strong coupling and large N_c, by replacing the D4-brane with the corresponding classical supergravity theory solution (black-4-brane) and considering an open string theory on the D8-$\overline{\text{D8}}$ brane put as a probe (probe approximation), we obtain an effective theory for QCD analysis. In classical supergravity theory, the decoupling limit corresponds to the vicinity of the black-4-brane horizon, then the analysis of massless QCD can be realized by examining the open string theory on the D8 brane considering the

Table 2.1 The brane configuration in the Sakai-Sugimoto model

	0	1	2	3	4	5	6	7	8	9
D4	$\circ$	$\circ$	$\circ$	$\circ$	$\circ$	$\times$	$\times$	$\times$	$\times$	$\times$
D8	$\circ$	$\circ$	$\circ$	$\circ$	$\times$	$\circ$	$\circ$	$\circ$	$\circ$	$\circ$
$\overline{\text{D8}}$	$\circ$	$\circ$	$\circ$	$\circ$	$\times$	$\circ$	$\circ$	$\circ$	$\circ$	$\circ$

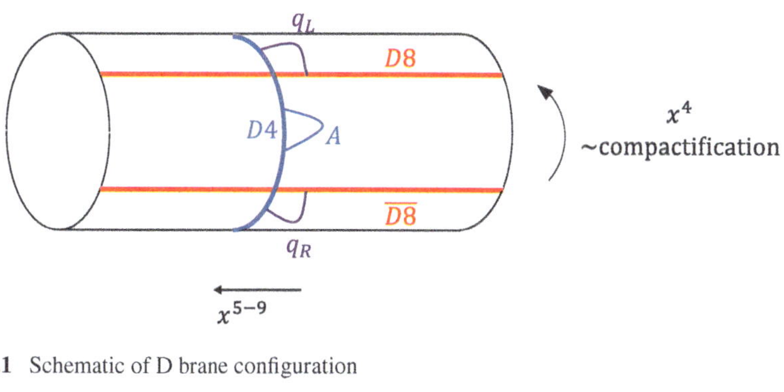

Fig. 2.1 Schematic of D brane configuration

induced metric in this background space-time. From this theory, the effective theory of mesons is obtained in which the baryon is analyzed as its soliton solution. The replacement of the black-4-brane shows the existence of a solution such that the D8, $\overline{\text{D8}}$ brane connects, which can be interpreted as a geometric realization of the spontaneously chiral symmetry breaking.

By placing the D4 brane at the origin of $\mathbb{R}^5$ labeled by x^5–x^9, the brane configuration of the Sakai-Sugimoto model is invariant to the SO(5) rotation that rotates the coordinate system of the space manifold $\mathbb{R}^5$. Since this symmetry does not exist in QCD, the fields appearing in the theory must be singlet for the SO(5) symmetry.

2.2.2 The Metric of the Black 4-Brane

For the analysis of QCD, we need a holographic dual of the described above brane configuration. Treating D8-$\overline{\text{D8}}$ as a probe, we are now interested in the near-horizon of the holographic dual of the D4 brane with one compactified direction, which corresponds to QCD. Similar to Appendix 2.5.2, the black 4-brane solution near the horizon is given by

$$ds^2 = \left(\frac{U}{R}\right)^{3/2}\left(\eta_{\mu\nu}dx^\mu dx^\nu + f(U)(dx^4)^2\right) + \left(\frac{R}{U}\right)^{3/2}\left(\frac{dU^2}{f(U)} + U^2 d\Omega_4^2\right) \tag{2.2.1}$$

$$e^\phi = g_s\left(\frac{U}{R}\right)^{3/4}, \quad F_4 = dC_3 = \frac{2\pi N_c}{V_4}\epsilon_4, \quad f(U) = 1 - \frac{U_{KK}^3}{U^3}, \tag{2.2.2}$$

with the Minkowski metric as $\eta_{\mu\nu} = (-1, 1, 1, 1)$ [6]. Here ϕ is the dilaton field, C_3 the RR 3-form, and F_4 its field strength. This ten-dimensional spacetime consists of the manifold $\mathbb{R}^4 \times [0, \infty) \times S^4$, where U labels the radial direction orthogonal to S^4 and $U = U_{KK}$ represents the horizon in the black 4-brane background spacetime [6]. The S^4 is labeled by Ω_4, and $V_4 = 8\pi^2/3$ and ϵ_4 are the volume and volume form of S^4, respectively. In the same way as for Appendix 2.5.2, the constant R is expressed by using the string coupling constant g_s and string length l_s as follows;

$$R^3 = \pi g_s N_c l_s^3. \tag{2.2.3}$$

To decouple the redundant fermions in this metric, x^4 is compactified to S^1. Consider the behavior of $f(U) = 1 - U_{KK}^3/U^3$ in front of this $(dx^4)^2$. For $U > U_{KK}$, as U increases, the radius of the circle around which x^4 is wrapped also increases. On the other hand, when $U = U_{KK}$, this circle vanishes, which means that at $U \leq U_{KK}$, spacetime no longer exists. Since chiral symmetry is realized as a global symmetry on D8 and $\overline{\text{D8}}$, the geometry in which D8 and $\overline{\text{D8}}$ are connected to the

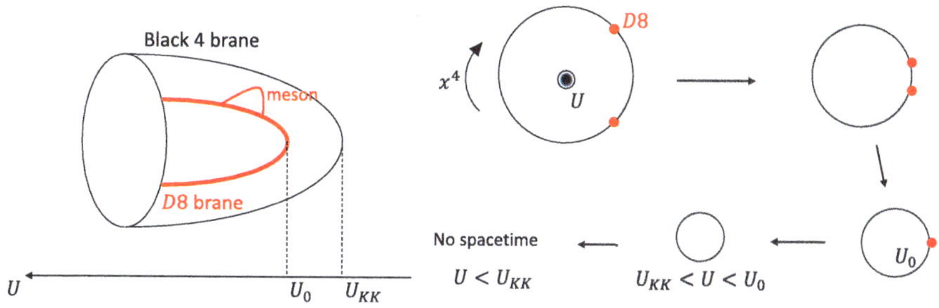

Fig. 2.2 Geometrical realization of the spontaneously chiral symmetry breaking

black 4-brane background spacetime represents the spontaneously chiral symmetry breaking (Fig. 2.2).

2.2.3 *Conditions to Be Satisfied by Parameters on the YM Theory Side*

Let us summarize the conditions for allowing the analysis of massless QCD by the gravity theory described above. In the following, the restrictions on the parameters R, U_{KK}, g_s, etc. in string theory are rewritten as conditions on parameters on the YM theory side. Furthermore, we find that all these constants can be expressed in the dimension M_{KK}, which makes the expression of the action very simple. This is very useful for the analysis in this thesis [2].

The relation between the YM coupling constant g_{YM} and the parameters of string theory can be written as follows by reading from the effective action of D4 brane with S^1 compactified direction (like Appendix 2.5.1);

$$g_{YM}^2 = \frac{g_s}{T_4(2\pi\alpha')^2\delta\tau} = \frac{(2\pi)^2 g_s l_s}{\delta\tau} \tag{2.2.4}$$

$$T_p = \frac{1}{(2\pi)^p l_s^{p+1}}, \quad \alpha' = l_s^2, \tag{2.2.5}$$

with tension for Dp-brane T_p.

Also, although there seems to be singularity in $U = U_{KK}$ in (2.2.1), the compactification of x^4 into S^1 naturally introduces periodicity in this axis by one rotation of S^1. By choosing this periodicity well, singularity can be avoided (see Appendix 2.5.4). With δx^4 for this periodic, singularity can be eliminated by choosing

$$\delta x^4 = \frac{4\pi}{3}\frac{R^{3/2}}{U_{KK}^{1/2}}. \tag{2.2.6}$$

In terms of the compactification radius $1/M_{KK}$, we are written as

$$\delta x^4 = 2\pi \frac{1}{M_{KK}}, \tag{2.2.7}$$

so there is a relation

$$M_{KK} = \frac{3}{2}\frac{U_{KK}^{1/2}}{R^{3/2}} \tag{2.2.8}$$

between M_{KK}, U_{KK} and R.

From the above, we see that the parameters on the string theory side and those on the YM theory side are related to

$$\begin{aligned} R^3 &= \frac{1}{2}\frac{g_{YM}^2 N_c l_s^2}{M_{KK}}, \quad U_{KK} = \frac{2}{9} g_{YM}^2 N_c M_{KK} l_s^2 \\ g_s &= \frac{1}{2\pi}\frac{g_{YM}^2}{M_{KK} l_s}. \end{aligned} \tag{2.2.9}$$

We now consider the conditions on the YM theory side for the validity of the description by the black 4-brane of the D4-brane. As mentioned in the previous section, for the description by classical supergravity theory to be valid, the string length must be sufficiently smaller than the considering energy scale, and the quantum corrections of the supergravity theory must be suppressed.

The curvature of the black-4-brane spacetime is $\mathcal{R} \sim \sqrt{\lambda} l_s$, which imposes the condition that l_s is smaller than the curvature $\mathcal{R}$, the typical length scale of spacetime. Thus, it is attributed the following conditions for 't Hooft coupling λ;

$$\frac{l_s}{\mathcal{R}} = \lambda^{-1/2} \ll 1 \quad \rightarrow \quad \lambda \gg 1. \tag{2.2.10}$$

The effective coupling constant e^{ϕ} of the string should be small enough for the loop correction of the closed string to be negligible, which is given by (2.2.2) is

$$e^{\phi} = g_s \left(\frac{U_{KK}}{R}\right)^{3/4} \left(\frac{U}{U_{KK}}\right)^{3/4} = \frac{g_{YM}^2 \lambda}{3\sqrt{3}\pi} \left(\frac{U}{U_{KK}}\right)^{3/4}, \tag{2.2.11}$$

with the relations (2.2.9). Since we are now considering a near-horizon neighborhood, $U/U_{KK} \sim 1$, the condition $e^{\phi} \ll 1$ is rewritten as

$$g_{YM}^4 \lambda \ll 1. \tag{2.2.12}$$

If we express the above requirements in terms of parameters on the YM theory side, we get

$$g_{YM}^2 \ll \frac{1}{\lambda} \ll 1. \tag{2.2.13}$$

This condition is realized by $g_{YM}^2 \to 0$ and $N_c \to \infty$, by taking λ finite yet large values, which shows that this massless QCD is a strongly coupled gauge theory with large N_c.

Furthermore, it can be seen that the string length l_s does not appear explicitly in the action $S_{\mathrm{D8}}^{\mathrm{DBI}}$ (2.3.5) that we analyze and derive later in this thesis. Since $U_{KK}^{1/2} R^{3/2} \gg l_s^2$ is required with $\lambda \gg 1$ as mentioned above, it is possible to set

$$l_s^2 = \frac{9}{2}\lambda^{-1} M_{KK}^{-2} \tag{2.2.14}$$

with a value of the same dimension that is small enough without losing generality. Using this and the relation (2.2.9), we can express them as M_{KK}, and if we set $M_{KK} = 1$, then we find that we can set them as

$$M_{KK} = 1, \quad R^3 = \frac{9}{4}, \quad U_{KK} = 1. \tag{2.2.15}$$

They can be easily recovered in the final dimensional analysis.

2.3 Mesons in the Sakai-Sugimoto Model

We can perform the study of mesons by examining the open string theory on the D8 brane, which is set as a probe in the black 4-brane. Figure 2.3 shows that the meson of the Sakai-Sugimoto model is an open string with both ends on the D8 brane. The open string with one end on the D4 brane and the other on the D8 brane corresponds to a quark. The distinction between quark and anti-quark is defined by the orientation of the strings. When the D4 brane is replaced by the black 4 brane, the two strings are attached and are interpreted as a meson, as shown in the lower right picture in Fig. 2.3. At this time, the definition of quark and anti-quark orientation works well.

A high-dimensional meson field appears as the massless mode of the open string on the D8 brane. By mode expansion of this meson field and dimensional reduction to four-dimensional space-time, we obtain the spectra of actual mesons. These meson spectra include not only pions but also vector and axial vector mesons and their resonance states. Considering that the actual nucleon resonance appears as a pole of scattering between the meson and the nucleon, it is reasonable to describe them as a meson-baryon multi-system. The dimensional reduction of the open string theory on the D8 brane leads to the action of the Skyrme model with scalar and vector mesons, which has already been studied in the framework of hidden local symmetry [7]. If we consider the skyrmion as a nucleon, it is indeed nothing but a meson-baryon composite system.

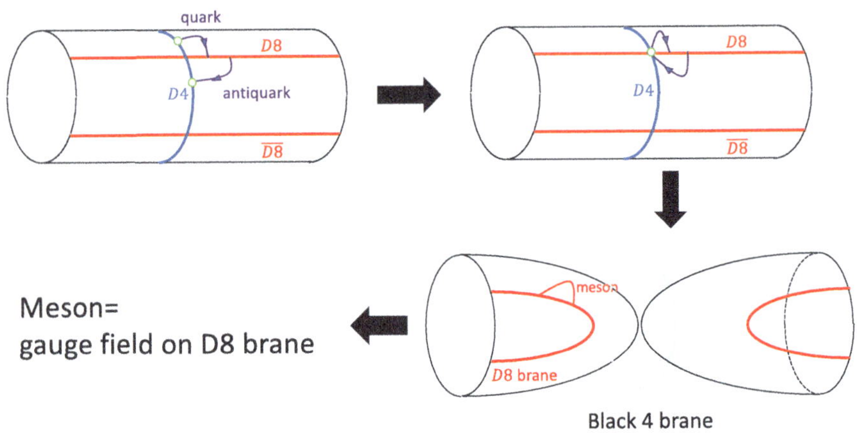

Fig. 2.3 Mesons in the Sakai-Sugimoto model

2.3.1 *Open String Theory on the D8 Brane*

Now consider the gauge field on the D8 brane placed in the black 4-brane background space-time. The gauge field on the D8 brane has nine components: $(A_\mu(x^\nu, U, y^\alpha)$, $A_U(x^\mu, U, y^\alpha)$, $A_\alpha(x^\mu, U, y^\beta))$, $(\mu, \nu = 0, 1, 2, 3, \quad \alpha, \beta = 6, 7, 8, 9)$. As mentioned in Sect. 2.2.1, S^4 is labeled by y^α while the radial direction orthogonal to them is U. The brane configuration is invariant under the SO(5) transformation that rotates the coordinate of $\mathbb{R}^5$ labeled by (U, y^α), therefore the gauge field on the D8 brane is also required to be invariant to this transformation (Appendix 3.5).

In the Sakai-Sugimoto model, this invariance is realized by only utilizing the radial component among the gauge field components on $\mathbb{R}^5$ and ignoring the others (Fig. 2.4: In this figure, the rotation of $\mathbb{R}^5$ is depicted as a rotation of a two-dimensional plane).

Fig. 2.4 The SO(5) symmetric gauge fields in the Sakai-Sugimoto model

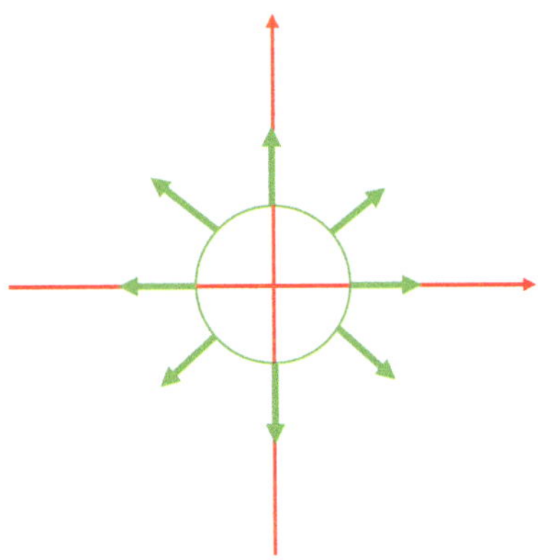

Namely, it means to consider the case where the gauge field on the D8 brane is $(A_\nu(x^\nu, U), A_U(x^\mu, U), A_\alpha = 0)$.

By using the gauge fields described above, the Dirac-Born-Infeld (DBI) action $S_{\rm D8}^{\rm DBI}$ on the D8 brane is written as

$$S_{\rm D8}^{\rm DBI} = T_8 \int d^9 x e^{-\phi} {\rm Tr}_f \sqrt{-\det(g_{\rm D8} + 2\pi\alpha' F)} \tag{2.3.1}$$

$$T_p = \frac{1}{(2\pi)^p l_s^{p+1}}, \quad e^\phi = g_s \left(\frac{U}{R}\right)^{3/4}, \quad \alpha' = l_s^2,$$

where l_s is string length, g_s string coupling constant, and ${\rm Tr}_f$ the trace for flavor space (Hereafter, unless otherwise noted, we consider the trace for the flavor.). The det is performed on a 9×9 matrix of space-time components, e.g., g_{MN} ($g_{\rm D8} = g_{MN} dx^M dx^N$) and $F_{MN} = \partial_M A_N - \partial_N A_M + [A_M, A_N]$. When we consider the D8 brane embedded in the black 4-brane, the induced metric on the D8 brane is written as [6, 8]

$$ds_{\rm D8}^2 = \left(\frac{U}{R}\right)^{3/2} \eta_{\mu\nu} dx^\mu dx^\nu + \left(\frac{R}{U}\right)^{3/2} f^{-1}(U) dU^2 + \left(\frac{R}{U}\right)^{3/2} U^2 d\Omega_4. \tag{2.3.2}$$

Using $\det(AB) = \det A \det B$ and $\det X = \det(e^{\ln X}) = e^{{\rm tr}(\ln X)}$ we obtain

$$\begin{aligned}\sqrt{-\det(g + 2\pi\alpha' F)} &= \sqrt{-\det(g)} (\det[1 + 2\pi\alpha' g^{-1} F])^{1/2} \\ &= \sqrt{-\det(g)} \exp\left[\frac{1}{2} {\rm tr} \ln(1 + 2\pi\alpha' g^{-1} F)\right] \\ &= \sqrt{-\det(g)} \exp\left[\frac{1}{2} {\rm tr}\left(2\pi\alpha' g^{-1} F - \frac{1}{2!}(2\pi\alpha' g^{-1} F)^2 + \mathcal{O}(F^4)\right)\right] \\ &= f^{-1/2}\left(\frac{R}{U}\right)^{3/4} U^4 \left(\frac{1}{2} + \frac{1}{4}(2\pi\alpha')^2 g^{MN} g^{PQ} F_{MP} F_{NQ} + \mathcal{O}(F^4)\right)\end{aligned} \tag{2.3.3}$$

Then, substituting $A_\alpha = 0$ and $e^\phi = g_s (U/R)^{3/4}$ into $S_{\rm D8}^{\rm DBI}$, we have the $S_{\rm D8}^{\rm DBI}$ as follow;

$$\begin{aligned}S_{\rm D8}^{\rm DBI} = T_8 g_s^{-1} (2\pi\alpha')^2 &\int d^4 x dU d\Omega_4 \left(\frac{R}{U}\right)^{3/4} U^4 f^{-1/2}(U) \\ &\times {\rm Tr}_f \left(\frac{1}{4} g^{\mu\nu} g^{\rho\sigma} F_{\mu\rho} F_{\nu\sigma} + \frac{1}{2} g^{\mu\nu} g^{UU} F_{\mu U} F_{\nu U}\right) \\ &+ {\rm const} + \mathcal{O}(F^4).\end{aligned} \tag{2.3.4}$$

Furthermore, we perform the following replacements, $g^{\mu\nu} = (U/R)^{-3/2} \eta^{\mu\nu}$, $g^{UU} = (R/U)^{-3/2} f(U)$ and

$$
\begin{aligned}
U &= U_{KK} u \\
u^3 &= 1 + z^2, \quad \frac{du}{dz} = \frac{2}{3} u^{-1/2} f^{1/2},
\end{aligned}
$$

then we finally obtain

$$
S^{\mathrm{DBI}}_{\mathrm{D8}} = \kappa \int d^4 x dz \mathrm{Tr}_f \left(\frac{1}{4} K^{-1/3} \eta^{\mu\nu} \eta^{\rho\sigma} F_{\mu\rho} F_{\nu\sigma} + \frac{1}{2} K \eta^{\mu\nu} F_{\mu z} F_{\nu z} \right) \tag{2.3.5}
$$

$$
\kappa = \frac{9}{4} (2\pi\alpha')^3 T_8 V_4 g_s^{-1} = \frac{\lambda N_c}{108\pi^3}, \tag{2.3.6}
$$

with $u(z)^3 = K(z) = 1 + z^2$ and (2.2.15). Here, because all fields does not depend on y^α, we integrate as follow;

$$
\int d\Omega_4 = V_4. \tag{2.3.7}
$$

2.3.2 Mode Expansion

To obtain the actual meson field, we employ a mode expansion of the five-dimensional gauge fields by a complete set of the function of z and obtain four-dimensional gauge fields [1]. It is arbitrary how to choose a complete set; we take the one to diagonalize the kinetic and the mass term in the four-dimensional space-time. Here we discuss how to find such a complete set systematically based on the discussion in Ref. [9], even though such a complete set is easily found in the original action of the Sakai-Sugimoto model. The key point is to impose an on-shell condition.

By ignoring interaction terms, the equations of motion are

$$
A_\mu \ : \ \partial_\nu \partial_\nu A_\mu - \partial_\mu \partial_\nu A_\nu + K^{1/3} \partial_z (K \partial_z A_\mu) - K^{1/3} \partial_z (K \partial_\mu A_z) = 0 \tag{2.3.8}
$$

$$
A_z \ : \ \partial_\mu \partial_\mu A_z - \partial_z \partial_\mu A_\mu = 0. \tag{2.3.9}
$$

Here, the gauge fields are Fourier transformed into momentum space

$$
A_\mu(x^\nu, z) = \int d^4 p \epsilon_\mu(p^\nu, z) e^{ip\cdot x} \tag{2.3.10}
$$

$$
A_z(x^\mu, z) = -i \int d^4 p \epsilon_z(p^\mu, z) e^{ip\cdot x}, \tag{2.3.11}
$$

to obtain

$$
- p^2 \epsilon_\mu + p_\mu p \cdot \epsilon + K^{1/3} \partial_z [K(\partial_z \epsilon_\mu - p_\mu \epsilon_z)] = 0 \tag{2.3.12}
$$

$$
- p^2 \epsilon_z + \partial_z (p \cdot \epsilon) = 0, \tag{2.3.13}
$$

where the polarization vector is redefined as

$$\tilde{\epsilon}_\mu = \epsilon_\mu - p_\mu \int dz \epsilon_z. \tag{2.3.14}$$

This redefinition corresponds to a gauge fixing, the $A_z = 0$ gauge, in position space. Then, equations of motion become

$$-p^2\tilde{\epsilon}_\mu + p_\mu p \cdot \tilde{\epsilon} + K^{1/3}\partial_z(K\partial_z\tilde{\epsilon}_\mu) = 0 \tag{2.3.15}$$
$$\partial_z(p \cdot \tilde{\epsilon}) = 0. \tag{2.3.16}$$

(a) Transverse mode ($p \cdot \tilde{\epsilon} = 0$)

In the case of the transverse mode, with the on-shell condition ($-p^2 = (m_n^V)^2$), the equation of motion (2.3.15) is

$$(m_n^V)^2\tilde{\epsilon}_\mu + K^{1/3}\partial_z(K\partial_z\tilde{\epsilon}_\mu) = 0, \tag{2.3.17}$$

while (2.3.16) is trivially satisfied. By performing a mode expansion $\tilde{\epsilon}_\mu(p^\mu, z) = \psi_n(z)X_\mu^n(x^\mu)$ and transferring it back to position space, the EOM becomes

$$(m_n^V)^2\psi_n(z) + K^{1/3}\partial_z(K\partial_z\psi_n(z)) = 0, \tag{2.3.18}$$

which is the eigenvalue equation that must be satisfied by the complete set $\psi_n(z)$ such that the kinetic term and the mass term are diagonalized. The mode expansion performed here is rewritten as follows,

$$\begin{aligned} A_\mu(x^\nu, z) &= \int d^4p\tilde{\epsilon}_\mu(p^\mu, z)e^{p\cdot x} = \psi_n(z)\int d^4pX_\mu^n(p)e^{ip\cdot x} \\ &= \sum_{n\geq 1}\psi_n(z)X_\mu^n(x^\nu) \end{aligned} \tag{2.3.19}$$

$$A_z(x^\mu, z) = 0. \tag{2.3.20}$$

(b) Longitudinal mode ($\tilde{\epsilon}_\mu = p_\mu\tilde{\epsilon}$)

In the case of the longitudinal mode, the equations of motion become

$$\partial_z(K\partial_z\tilde{\epsilon}) = 0 \tag{2.3.21}$$
$$(m_n^S)^2\partial_z(\tilde{\epsilon}) = 0. \tag{2.3.22}$$

Solving the above equation, we obtain

$$\partial_z\tilde{\epsilon} = C_1(p^\mu)K^{-1} \tag{2.3.23}$$
$$\tilde{\epsilon} = C_1(p^\mu)\arctan z + C_2(p^\mu) \tag{2.3.24}$$

with an arbitrary p^μ function $C_1(p^\mu)$. Then, the lower equation means $(m_n^S)^2 = 0$. The mode expansion is written as

$$\begin{aligned} A_\mu(x^\mu, z) &= \int d^4 p p_\mu \tilde{\epsilon}(p^\nu, z) e^{ip\cdot x} \\ &= \arctan z(-i)\partial_\mu\Big(\int d^4 p C_1(p^\nu) e^{ip\cdot x}\Big) + C_2 \\ &= C_1 \arctan z \partial_\mu Y^0(x^\nu) + C_2 \qquad (2.3.25) \\ A_z(x^\mu, z) &= 0, \qquad (2.3.26) \end{aligned}$$

where $C_{1,2}$ are determined by normalization and $C_2 = 0$.

From the above, together with transverse and longitudinal mode, we obtain the mode expansion

$$A_\mu(x^\mu, z) = \sum_{n\geq 1} \psi_n(z) X_\mu^n(x^\nu) + C_1 \arctan z \partial_\mu Y^0(x^\nu) \qquad (2.3.27)$$

$$A_z(x^\mu, z) = 0. \qquad (2.3.28)$$

if we perform a gauge transformation, we get

$$A_\mu(x^\mu, z) = \sum_{n\geq 1} \psi_n(z) X_\mu^n(x^\nu) + C_1 \arctan z \partial_\mu Y^0(x^\nu) \qquad (2.3.29)$$

$$A_z(x^\mu, z) = 0, \qquad (2.3.30)$$

where, in the Ref. [1], $X_\mu^n(x^\nu)$ and $Y^n(x^\mu)$ are denoted as $B_\mu^n(x^\nu)$ and $\varphi^n(x^\mu)$ respectively. The above discussion also applies to the case where a mass term is added to the 5-dimensional YM action.

Without the interaction term and substituting the mode expansion (2.3.29) into (2.3.5), we obtain

$$F_{\mu\nu} = (\partial_\mu X_\nu^n - \partial_\nu X_\mu^n) \sum_{n\geq 1} \psi_n(z) \qquad (2.3.31)$$

$$F_{\mu z} = -X_\mu^n \sum_{n\geq 1} \partial_z \psi_n(z) - \partial_\mu Y^0 C_1 K^{-1}. \qquad (2.3.32)$$

Using the eigenvalue equation (2.3.18) and the normalization condition

$$\kappa \int dz K^{-1/3} \psi_n \psi_m = \delta_{nm} \qquad (2.3.33)$$

$$\kappa \int dz K C_1^2 (K^{-1})^2 = 1, \qquad (2.3.34)$$

we get a 4-dimensional meson effective action;

$$S_{\mathrm{D8}}^{\mathrm{DBI}} = -\int d^4x \mathrm{Tr}_f\Big[\frac{1}{2}\eta^{\mu\nu}\partial_\mu Y^0\partial_\nu Y^0 + \sum_{n\geq 1}\Big(\frac{1}{4}\eta^{\mu\nu}\eta^{\rho\sigma}F^n_{\mu\rho}F^n_{\nu\sigma} + \frac{1}{2}(m^V_n)^2\eta^{\mu\nu}X^n_\mu X^n_\nu\Big)\Big] \tag{2.3.35}$$

with $F^n_{\mu\nu} = \partial_\mu X^n_\nu - \partial_\nu X^n_\mu$. Let me comment on what can be learned from the above discussion. First, we see that Y^0 is a massless scalar field. It will be easy to guess that this corresponds to a pion field. Second, B^n_μ is a vector field, whose mass $(m^V_n)^2$ should be determined by finding the eigenvalues of the eigenvalue equation (2.3.18). The vector and axial vector meson spectra are obtained in this way and are found to be in good agreement with experimental values. This is explained below.

2.3.3 *Parity and Charge Conjugation*

We have discussed how the spectrum of vector mesons is obtained, however, to identify the fields that correspond to actual mesons, it is necessary to investigate the transformational properties of these vector fields concerning the parity transformation P and the charge conjugate transformation C.

At first, the action is obviously invariant under the 5-dimensional proper Lorentz transformation $(t, x^1, x^2, x^3, z) \to (t, -x^1, -x^2, -x^3, -z)$. This transformation is nothing but a parity transformation in 4-dimensional theory after the z integral. Thus, we have the parity transformation

$$P\ :\ (t, x^1, x^2, x^3, z) \to (t, -x^1, -x^2, -x^3, -z) \tag{2.3.36}$$

in this 5-dimensional theory. Because this parity transformation P is a proper Lorentz transformation, the transformation property of the gauge field is

$$P\ :\ (A_0, A_1, A_2, A_3, A_z) \to (A_0, -A_1, -A_2, -A_3, -A_z), \tag{2.3.37}$$

namely,

$$P\ :\ (A_\mu, A_z) \to (A^\mu, -A_z). \tag{2.3.38}$$

Furthermore, because (2.3.18) is invariant with respect to $P : z \to -z$, $\psi_n(z)$ is either an even or an odd function. By solving the eigenvalue equation, we find

$$
\begin{aligned}
n = 1, &\quad \psi_1(z) \ : \ \text{even} \\
n = 2, &\quad \psi_2(z) \ : \ \text{odd} \\
n = 3, &\quad \psi_3(z) \ : \ \text{even} \\
&\quad \vdots
\end{aligned}
$$

and determine that $\phi_0 = K^{-1} = (1 + z^2)^{-1}$ is an even function. Considering this properties of the fields and (2.3.38), we find

$$
\begin{aligned}
n = 1, &\quad X^{(1)}_{\mu}(x^{\mu}) \to -X^{\mu(1)} \\
n = 2, &\quad X^{(2)}_{\mu}(x^{\mu}) \to +X^{\mu(2)} \\
n = 3, &\quad X^{(3)}_{\mu}(x^{\mu}) \to -X^{\mu(3)} \\
&\quad \vdots
\end{aligned}
$$

and we finally conclude

$$
\begin{aligned}
Y^0 &\ : \ \text{pseudo scalar particle} \\
X^1_{\mu} &\ : \ \text{vector particle} \\
X^2_{\mu} &\ : \ \text{pseudo vector particle} \\
X^3_{\mu} &\ : \ \text{vector particle} \\
&\quad \vdots
\end{aligned}
$$

with attention to the raising and lowering indices.

On the other hand, the charge conjugation C switches particles and antiparticles, which from the string theory viewpoint corresponds to reverse the orientation of the string. Therefore, $C : A \to -A^T$ corresponds to a charge conjugation. However, by this definition, we see that the DBI action is invariant under this transformation, but the CS term,

$$
\int C_3 \mathrm{Tr}_f F^3, \tag{2.3.39}
$$

changes its sign.

Therefore, let us define

$$C\ :\ (A, z) \to (-A^T, -z) \tag{2.3.40}$$

to keep the action invariant. Because we are concerned with charge conjugation in four-dimensional theory, the property of transformation for z is arbitrary as long as the integral value remains unchanged. Thus, the charge conjugation of the field is determined to be

$$C\ :\ A_\mu \to -A_\mu^T,\quad A_z \to A_z^T. \tag{2.3.41}$$

From the above, we determine

$$\begin{aligned} Y^0 &: 0^{-+} \\ X_\mu^1 &: 1^{--} \\ X_\mu^2 &: 1^{++} \\ X_\mu^3 &: 1^{--} \\ &\ \ \vdots \end{aligned}$$

with the J^{PC} of the fields in the four-dimensional theory. As expected, Y^0 is an NG boson, or pion, associated with spontaneous breaking of the chiral symmetry, and has zero mass in the case of the chiral limit. For (axial-)vector mesons, we also identify them as $X_\mu^1 : \rho$, $X_\mu^2 : a_1$, $X_\mu^3 : \rho'(1465)$, $X_\mu^4 : a_1'(1640)$, ...etc. Finally, in the next section, we will calculate the masses of these (axial-)vector mesons to verify that we reproduce the experimental values.

2.3.4 Spectra for (axial-)vector Mesons

To calculate the mass of a (axial-)vector meson, we just need to solve for Eq. (2.3.18). As this eigenvalue equation is solved numerically in Ref. [1], we will discuss their method of computation. The eigenvalue equations with their normalization conditions that must be solved are shown once again as follows;

$$-K^{1/3}\partial_z(K\partial_z\psi_n(z)) = (m_n^V)^2\psi_n(z) \tag{2.3.42}$$

$$\kappa\int dzK^{-1/3}\psi_n\psi_m = \delta_{nm}. \tag{2.3.43}$$

To solve the eigenvalue equation, we need to know the asymptotic behavior of $\psi_n(z)$. First, in order to avoid divergence of the integral of the normalization condition, the integrand should be such that it decreases more rapidly than z^{-1} at $z \to \infty$.

Therefore, the asymptotic behavior,

$$\psi_n(z) \sim \mathcal{O}(z^a), \quad a < -1/6, \quad (z \to \infty) \tag{2.3.44}$$

is required. Also, at large z, the eigenvalue equation becomes

$$z^2 \psi_n(z) + z\psi_n(z) = 0. \tag{2.3.45}$$

From these two observations, we know that the behavior of $\psi_n(z)$ at $z \to \infty$ is

$$\psi_n(z) \sim \mathcal{O}(z^{-1}). \tag{2.3.46}$$

Therefore, if we define a new function

$$\tilde{\psi}_n(z) = z\psi_n(z), \tag{2.3.47}$$

we can conclude that it behaves as

$$\tilde{\psi}_n(z) \sim \mathcal{O}(z^0), \quad (z \to \infty). \tag{2.3.48}$$

For the new function $\tilde{\psi}_n(z)$, we find a formal solution in the form of an infinite series. The eigenvalue equation for this new function is

$$K\partial_z^2\tilde{\psi}_n - \frac{2}{z}\partial_z\tilde{\psi}_n + \left(\frac{2}{z^2} + (m_n^V)^2 K^{-1/3}\right)\tilde{\psi}_n = 0. \tag{2.3.49}$$

A transformation into η for $z = e^\eta$ this equation become

$$\partial_\eta^2\tilde{\psi}_n + A\partial_\eta\tilde{\psi}_n + B\tilde{\psi}_n = 0 \tag{2.3.50}$$

$$A = -\frac{1+3e^{-2\eta}}{1+e^{-2\eta}} = \sum_{l=0}^{\infty} A_l e^{-\frac{2l}{3}\eta} \tag{2.3.51}$$

$$B = \frac{2e^{-2\eta}}{1+e^{-2\eta}} + \lambda_n e^{-\frac{2}{3}\eta}(1+e^{-2\eta})^{-4/3} = \sum_{l=0}^{\infty} B_l e^{-\frac{2l}{3}\eta}, \tag{2.3.52}$$

and we that A and B are given as infinite series of $e^{-\frac{2l}{3}\eta}$, with

$$A_0 = -1,\ A_1 = 0,\ A_2 = 0,\ A_3 = -2,\ A_4 = 0,\ A_5 = 0, \ldots. \tag{2.3.53}$$

$$B_0 = 0,\ B_1 = \lambda_n,\ B_2 = 0,\ B_3 = 2,\ B_4 = -\frac{4}{3}\lambda_n,\ B_5 = 0, \ldots. \tag{2.3.54}$$

From this, by expanding $\tilde{\psi}_n$ with

Table 2.2 The spectra for vector mesons

	ρ	a_1	ρ'	a_1'
Theory	[776]	1189	1607	2023
Experimental	776^{--}	1230^{++}	1465^{--}	1720^{++}

$$\tilde{\psi}_n = \sum_l \alpha_l e^{-\frac{2l}{3}\eta}, \tag{2.3.55}$$

the α_l are obtained by a recurrence formula. Substituting this series expansion $\tilde{\psi}_n$ for (2.3.50), we obtain the following recurrence formula as

$$\frac{4l^2}{9}\alpha_l - \frac{2}{3}\sum_{m=1}^{l} m A_{l-m}\alpha_m + \sum_{m=0}^{l-1} B_{l-m}\alpha_m = 0. \tag{2.3.56}$$

By choosing $\alpha_0 = \pm 1$ appropriately and solving this recurrence formula sequentially, we obtain

$$\alpha_1 = -\frac{9}{10}(m_n^V)^2, \quad \alpha_2 = \frac{81}{280}(m_n^V)^4, \quad \alpha_3 = -\frac{1}{3} - \frac{27}{560}(m_n^V)^6, \ldots. \tag{2.3.57}$$

At large z, it is sufficient to truncate $e^{-\frac{2l}{3}\eta}$ at about $l = 3$. In the region of large z, using (2.3.55), the differential equation is solved toward the origin by the difference method. At the origin, shooting method with the boundary condition

$$\partial_z \psi_n(0) = 0 \quad or \quad \psi_n(0) = 0 \tag{2.3.58}$$

yields the result

$$(m_n^V)^2 = 0.67^{--}, \quad 1.6^{++}, \quad 2.9^{--}, \quad 4.5^{++}, \ldots. \tag{2.3.59}$$

If we restore M_{KK} by dimensional analysis and determine M_{KK} to reproduce the experimental value of 776 MeV for the mass ρ, we obtain Table 2.2. Here, we chose $\alpha_0 = 1$ for ρ and a_1, and $\alpha_0 = -1$ ρ' and a_1'. Also, M_{KK} was chosen to be 949 MeV.

2.4 Baryons in the Sakai-Sugimoto Model

In the previous section, we saw that the open string theory on D8 brane is a five-dimensional effective theory of mesons that includes also resonances. If we reduce this theory to four dimensions, we find that the YM term yields the Skyrme action, and the CS term yields the WZW term. Considering that Skyrmion is regarded as a baryon, this effective theory is interpreted as a meson-baryon composite system in four-dimensional space-time. Moreover, this fact is important for the explanation of realistic nucleon resonances.

In the Sakai-Sugimoto model, baryons are interpreted as D4 branes (baryon vertices) wrapped around S^4 [10, 11]. The open string theory on the D4 brane (baryon vertex) that characterizes the dynamics of this baryon is regarded as ADHM data. This ADHM data is interpreted as instantons on the D8 brane. The relation between the instanton and the Skyrmion is written as

$$U(x^{\mu}) = \mathrm{P}\exp\Big(-\int_{-\infty}^{\infty} dz' A_z(x^{\mu}, z')\Big), \tag{2.4.1}$$

where $U(x^{\mu})$ is the chiral field [12]. Thus, in the Sakai-Sugimoto model, the soliton interpretation of baryons is justified from the standpoint of superstring theory.

Studies of baryons by the Sakai-Sugimoto model have been done by Refs. [13, 14]. In this thesis, we use the latter approach which is much easier to analyze. In this approach, baryons are treated and analyzed as instantons in a 4-dimensional space in a 5-dimensional space-time.

2.4.1 *Classical Solutions*

The first step is to find a static solution that is independent of time. The time dependence is introduced by assigning it to the collective coordinates. This is a method called the moduli space approximation method, which is used when considering the case of slow-moving solitons [15, 16].

CS Term

At first, we introduce the CS term, which has been neglected until now. The effective action of Dp branes is described by DBI action and CS term [17]. The CS term is

$$S^{\mathrm{CS}}_{\mathrm{D8}} = \mu_p \int_{\mathrm{D8}} \sum_q C_q \wedge \mathrm{Tr}(e^{2\pi\alpha' F_2 + B_2}) \tag{2.4.2}$$

$$\mu_p = T_p = \frac{1}{(2\pi)^p l_s^{p+1}}, \tag{2.4.3}$$

where C_p is the RR p-form field, B_2 the KR field, and $\wedge$ the wedge product. From the properties required for the black 4-brane solution, we have $B_2 = 0$, $C_k = 0$ $(k \neq 7-4)$, and the sum remains valued only at $q = 3$. Since this integral is in nine-dimensional spacetime, only the 9-form survives. Performing the series expansion, there is only one term that is the 9-form together with the 3-form C_3, then, the CS term is

$$S_{\text{D8}}^{\text{CS}} \propto \int_{\text{D8}} C_3 \wedge \text{Tr}(F_2)^3. \tag{2.4.4}$$

Since $(F_2)^3$ is a 6-form, we see that it is zero as long as we consider the five components of (A_μ, A_z).

With

$$(F_2(A))^3 = d\omega_5(A) \tag{2.4.5}$$

$$\omega_5(A) = \text{Tr}\Big(AF^2 - \frac{i}{2}A^3F - \frac{1}{10}A^5\Big), \tag{2.4.6}$$

if we remind that it was

$$F_4 = dC_3 = \frac{2\pi N_c}{V_4}\epsilon_4 \tag{2.4.7}$$

in the black 4 brane solution in (2.2.1), the (2.4.4) is computed as

$$\begin{aligned}\int_{\text{D8}} C_3 \wedge d\omega_5 &= \int_{\mathbb{R}^4\times[0,\infty)\times S^4} dC_3 \wedge \omega_5 \\ &= \frac{2\pi N_c}{V_4}\int_{S^4}\epsilon_4 \wedge \int_{\mathbb{R}^4\times[0,\infty)}\omega_5\end{aligned} \tag{2.4.8}$$

(it is non-trivial to ignore the surface term), and using

$$\int_{S^4} F_4 = 2\pi, \tag{2.4.9}$$

we have the final expression of the CS term

$$S_{\text{D8}}^{\text{CS}} = \frac{N_c}{24\pi^2}\int_{\mathbb{R}^4\times[0,\infty)}\omega_5(A). \tag{2.4.10}$$

Action

The action used in the analysis of baryons is

$$\begin{aligned}S_{\text{D8}} &= S^{\text{DBI}} + S^{\text{CS}} \\ &= -\kappa\int d^4x dz \text{Tr}_f\Big(\frac{1}{4}K(z)^{-1/3}\mathcal{F}_{\mu\nu}^2 + \frac{1}{2}K(z)\mathcal{F}_{\mu z}^2\Big)\end{aligned}$$

$$+\frac{N_c}{24\pi^2}\int_{\mathbb{R}^4\times[0,\infty)}\omega_5(\mathcal{A}) \tag{2.4.11}$$

$$\kappa=\frac{\lambda N_c}{108\pi^3},\quad K(z)=1+z^2 \tag{2.4.12}$$

$$\omega_5(\mathcal{A})=\mathrm{Tr}\Big(\mathcal{A}\mathcal{F}^2-\frac{i}{2}\mathcal{A}^3\mathcal{F}-\frac{1}{10}\mathcal{A}^5\Big), \tag{2.4.13}$$

omitting the notation D8 and using the shorthand notation $\eta^{\mu\nu}\eta^{\rho\sigma}\mathcal{F}_{\mu\rho}\mathcal{F}_{\mu\sigma}=\mathcal{F}^2_{\mu\nu}$. The Hermitian conjugate $U(N_f)$ gauge field is denoted by $\mathcal{A}$ ($\mathcal{A}^\dagger=\mathcal{A}$). Sometimes it is more convenient to use the anti-Hermitian conjugate gauge field, so in such cases, after a word of caution, we denote it by $\mathcal{A}\to i\mathcal{A}$ ($\mathcal{A}^\dagger=-\mathcal{A}$). When there is no subscript, it is assumed that the notation is in differential form. This U(1) term physically corresponds to an ω meson (or ϕ meson). Taking into account that, in the Skyrme model, the ω meson is responsible for stabilizing the soliton solution, this term is expected to play an important role in the present case as well. The field strength is a term in differential form, denoted

$$\mathcal{F}=d\mathcal{A}+i\mathcal{A}\wedge\mathcal{A}. \tag{2.4.14}$$

d is the exterior derivative and $\wedge$ is the wedge product.

The $U(N_f)$ gauge field is decomposed into and $U(1)$ and written as

$$\mathcal{A}=A^at^a+\frac{1}{\sqrt{2N_f}}\hat{A}I, \tag{2.4.15}$$

where I is the unit matrix of $N_f\times N_f$. The coefficients of the $U(1)$ term are chosen to be

$$\mathrm{Tr}_f\Big(\frac{1}{\sqrt{2N_f}}I\frac{1}{\sqrt{2N_f}}I\Big)=\frac{1}{2N_f}\mathrm{Tr}_f(I)=\frac{1}{2}, \tag{2.4.16}$$

aligned with the normalization of

$$\mathrm{Tr}_f(t^at^b)=\frac{1}{2}\delta^{ab}. \tag{2.4.17}$$

Then, the action is written as

$$S^{\mathrm{DBI}} = -\frac{\lambda N_c}{108\pi^3}\int d^4x dz \mathrm{Tr}_f\Big(\frac{1}{4}K^{-1/3}F_{\mu\nu}^2 + \frac{1}{2}KF_{\mu z}^2\Big)$$
$$-\frac{1}{2}\kappa\int d^4x dz\Big(\frac{1}{4}K^{-1/3}\hat{F}_{\mu\nu}^2 + \frac{1}{2}K\hat{F}_{\mu z}^2\Big) \tag{2.4.18}$$
$$S^{\mathrm{CS}} = \frac{N_c}{24\pi^2}\int\Big[\omega_5(A) + \frac{3}{\sqrt{2N_f}}\hat{A}\mathrm{Tr}_f F^2 + \frac{1}{2\sqrt{2N_f}}\hat{A}\hat{F}^2$$
$$+\frac{1}{\sqrt{2N_f}}d\Big(\hat{A}\mathrm{Tr}_f(2FA - \frac{i}{2}A^3)\Big)\Big], \tag{2.4.19}$$

where (2.4.16) is used. To obtain the above expression, which separates the U(1) part of the CS term, requires a complicated calculation, which is shown in Appendix 2.7.1. The calculations so far are valid for arbitrary N_f.

In the rest of this section we will mainly discuss the case $N_f = 2$. In this case, the identity (Appendix 2.7.2) is

$$\omega_5(A) = 0, \tag{2.4.20}$$

and the action is

$$S^{\mathrm{DBI}} = -\frac{\lambda N_c}{108\pi^3}\int d^4x dz \mathrm{Tr}_f\Big(\frac{1}{4}K^{-1/3}F_{\mu\nu}^2 + KF_{\mu z}^2\Big) \tag{2.4.21}$$
$$S^{\mathrm{CS}} = \frac{N_c}{24\pi^2}\epsilon_{MNPQ}\int d^4x dz\Big[\frac{3}{8}\hat{A}_0\mathrm{Tr}_f(F_{MN}F_{PQ}) - \frac{3}{2}\hat{A}_M\mathrm{Tr}_f(\partial_0 A_N F_{PQ})$$
$$+\frac{3}{4}\hat{F}_{MN}\mathrm{Tr}_f(A_0 F_{PQ}) + \frac{1}{16}\hat{A}_0\hat{F}_{MN}\hat{F}_{PQ} - \frac{1}{4}\hat{A}_M\hat{F}_{0N}\hat{F}_{PQ} + \text{(total derivative terms)}\Big], \tag{2.4.22}$$

where $M, N, P, Q = 1, 2, 3, z$ and $\epsilon^{0123z} = \epsilon_{123z} = +1$ (see Appendix 2.7.3 for the calculation of the CS term). In the following parts of this section, $M, N, P, Q = 1, 2, 3, z$ unless otherwise noted.

The Stability of the Instanton Solution

Because the direction z is curved and the time component is coupled, It is generally difficult to find the classical solution of the obtained action (2.4.24). If we now recall $\lambda \gg 1$, we will expect the $1/\lambda$ expansion to work. As we will soon show, this expansion works well to find the classical solution analytically, keeping up to the next-leading of the $1/\lambda$ expansion. If we also drop the next-leading of the $1/\lambda$ expansion, the instanton shrinks, and there is no stable solution.

First, by using

$$\begin{aligned}\epsilon_{ijk}F_{jk}\epsilon_{imn}F_{mn} &= (\delta_{jm}\delta_{kn} - \delta_{jn}\delta_{km})F_{jk}F_{mn}\\ &= F_{jk}F_{jk} - F_{jk}F_{kj} = 2F_{jk}F_{jk},\end{aligned} \tag{2.4.23}$$

the lower bound of energy is found to be

$$\begin{aligned}E(\rho) &= \frac{\kappa}{4}\int d^4xdz\Big(\frac{1}{2}K^{-1/3}(\epsilon_{ijk}F^a_{jk})^2 + K(F^a_{iz})^2\Big)\\ &\geq \frac{\kappa}{2}\int d^4xdz\Big(\sqrt{K(z)^{-1/3}K(z)}|\epsilon_{ijk}F^a_{jk}F^a_{iz}|\Big)\\ &\geq \Big|\frac{\kappa}{2}\int d^4xdz\epsilon_{ijk}F^a_{jk}F^a_{iz}\Big|\\ &= 8\pi^2\kappa|N_B|,\end{aligned} \tag{2.4.24}$$

where the second line of deformation used Schwartz's inequality, and the last expression used

$$\begin{aligned}N_B &= -\frac{1}{32\pi^2}\int d^3xdz\epsilon_{MNPQ}\mathrm{tr}(F_{MN}F_{PQ})\\ &= -\frac{1}{16\pi^2}\int d^3xdz\epsilon_{ijk}F^a_{jk}F^a_{iz}.\end{aligned} \tag{2.4.25}$$

The equality condition is

$$\sqrt{K(z)^{-1/3}}\epsilon_{ijk}F^a_{jk} = \sqrt{K(z)}F^a_{iz}, \tag{2.4.26}$$

and this is satisfied at $\rho = 0$, so if we drop the next leading of the $1/\lambda$ expansion, the instanton shrinks.

$1/\lambda$ *Expansion*

We have shown that if we drop the next leading of the $1/\lambda$ expansion, there is no stable solution. If we include up to the next leading, we no longer ignore the CS term. Then, we expect the solution to be stable as follows [14]. If we focus on the first term of the CS term (2.4.22), we find

$$\epsilon_{MNPQ}\int d^4xdz\hat{A}_0\mathrm{Tr}_f(F_{MN}F_{PQ}). \tag{2.4.27}$$

This is the term that represents the Coulomb interaction in five dimensions, whose energy is known to behave as $1/\rho^2$. Therefore, it is expected that the instanton do not shrink because of this term. It has also been discussed that the soliton is stabilized by the ω meson, in the context of the Skyrme model [18, 19]. Since the U(1) part of the gauge field corresponds to the ω meson, it is expected to stabilize the instanton solution similarly to the Skyrme model.

The current approach considers the motion of the soliton in moduli space, so the spatial integration is performed at the end. Therefore, it is better to do a rescale so that the classical solution is easy to obtain. By choosing a good rescale, the equations of motion for each field can be separated to easily find the classical solution. Such a rescale is

$$\begin{aligned} x^M &\to \lambda^{-1/2} x^M, \quad \mathcal{A}_M \to \lambda^{1/2} \mathcal{A}_M \\ \mathcal{F}_{MN} &\to \lambda \mathcal{F}_{MN}, \quad \mathcal{F}_{0M} \to \lambda^{1/2} \mathcal{F}_{0M}. \end{aligned} \tag{2.4.28}$$

Under this rescale the action become

$$\begin{aligned} S^{DBI} &= -aN_c \int d^4xdz \mathrm{Tr}_f \Big[\frac{\lambda}{2} F_{MN}^2 + \Big(-\frac{z^2}{6} F_{ij}^2 + z^2 F_{iz}^2 - F_{0M}^2 \Big) + \mathcal{O}(\lambda^{-1}) \Big] \\ &\quad - \frac{1}{2} aN_c \int d^4xdz \Big[\frac{\lambda}{2} \hat{F}_{MN}^2 + \Big(-\frac{z^2}{6} \hat{F}_{ij}^2 + z^2 \hat{F}_{iz}^2 - \hat{F}_{0M}^2 \Big) + \mathcal{O}(\lambda^{-1}) \Big] \end{aligned} \tag{2.4.29}$$

$$\begin{aligned} S^{CS} &= \frac{N_c}{24\pi^2} \epsilon_{MNPQ} \int d^4xdz \Big[\frac{3}{8} \hat{A}_0 \mathrm{Tr}_f(F_{MN} F_{PQ}) - \frac{3}{2} \hat{A}_M \mathrm{Tr}_f(\partial_0 A_N F_{PQ}) \\ &\quad + \frac{3}{4} \hat{F}_{MN} \mathrm{Tr}_f(A_0 F_{PQ}) + \frac{1}{16} \hat{A}_0 \hat{F}_{MN} \hat{F}_{PQ} - \frac{1}{4} \hat{A}_M \hat{F}_{0N} \hat{F}_{PQ} + (\text{total derivatives}) \Big], \end{aligned} \tag{2.4.30}$$

with $i, j = 1, 2, 3$. Because of $K^{-1/3} \to (1 + (\lambda^{-1/2} z)^2)^{-1/3} \simeq 1 - \frac{1}{3} \lambda^{-1} z^2 + \mathcal{O}(\lambda^{-2})$, note that the effects of curved extra dimensions are sub-leading of $1/\lambda$ expansion. Because the 1-form $\mathcal{A}$ does not change under the rescale, the form of the CS term, which is a 5-form integral, also do not change.

Classical Solutions

We perform the $1/\lambda^0$ expansion and obtained a solution order by order. The action (2.4.29) gives the EOM;

$$A_0 : D_M F_{0M} + \frac{1}{64\pi^2 a} \epsilon_{MNPQ} \hat{F}_{MN} F_{PQ} + \mathcal{O}(\lambda^{-1}) = 0 \tag{2.4.31}$$

$$A_M : D_N F_{MN} + \mathcal{O}(\lambda^{-1}) = 0 \tag{2.4.32}$$

$$\hat{A}_0 : \partial_M \hat{F}_{0M} + \frac{1}{64\pi^2 a} \epsilon_{MNPQ} \Big[\mathrm{Tr}_f(F_{MN} F_{PQ}) + \frac{1}{2} \hat{F}_{MN} \hat{F}_{PQ} \Big] + \mathcal{O}(\lambda^{-1}) \tag{2.4.33}$$

$$\hat{A}_M : \partial_N \hat{F}_{MN} + \mathcal{O}(\lambda^{-1}) = 0. \tag{2.4.34}$$

In the leading order, the SU(2) and the U(1) part of the gauge field are obtained as follow. First, within the above EOM, the classical solution of (2.4.32) is called the instanton solution. The solution with topological number (baryon number) $B = 1$ is known as a BPST instanton [20] and is written as

$$A_M(x) = -if(\xi)g\partial_M g^{-1} \tag{2.4.35}$$

$$f(\xi) = \frac{\xi^2}{\xi^2+\rho^2}, \quad \xi = \sqrt{(\vec{x}-\vec{X})^2+(z-Z)^2} \tag{2.4.36}$$

$$g(x) = \frac{(z-Z)-i(\vec{x}-\vec{X})\cdot\boldsymbol{\tau}}{\xi}, \tag{2.4.37}$$

where $(\vec{X}, Z)$ represents the position of the instanton in (x^1, x^2, x^3, z) space and ρ is the size of the instanton. Next, (2.4.34) requires that $\hat{A}_M$ be pure gauge, which always is vanished by a gauge transformation, then we obtain

$$\hat{A}_M = 0. \tag{2.4.38}$$

In the next to leading order, by using the above two solutions, we obtain the time-components of the SU(2) and U(1) gauge field as follow. Gauss's law (2.4.31) becomes

$$D_M^2 A_0 = 0, \tag{2.4.39}$$

which is the solution to this equation obtained in the Appendix 2.7.6. Finally, by substituting the solution obtained so far into (2.4.41), we get

$$\partial_M^2 \hat{A}_0 + \frac{3}{\pi^2 a}\frac{\rho^4}{(\xi^2+\rho^2)^4} = 0. \tag{2.4.40}$$

Solving this equation gives the solution,

$$\hat{A}_0 = \frac{1}{8\pi^2 a}\frac{1}{\xi^2}\left(1-\frac{\rho^4}{(\xi^2+\rho^2)^2}\right). \tag{2.4.41}$$

The solution obtained from the above can be summarized as

$$A_0 = 0 \tag{2.4.42}$$

$$A_M(x) = -if(\xi)g\partial_M g^{-1} \tag{2.4.43}$$

$$\hat{A}_0 = \frac{1}{8\pi^2 a}\frac{1}{\xi^2}\left(1-\frac{\rho^4}{(\xi^2+\rho^2)^2}\right) \tag{2.4.44}$$

$$\hat{A}_M = 0, \tag{2.4.45}$$

which is an instanton solution with baryon number one.

Finally, we obtain the mass of the soliton for the static case. The action (2.4.29) has the form

$$S = S^{\mathrm{DBI}} + S^{\mathrm{CS}} = \int dt (E(t)_{\text{kinetic}} - E(t)_{\text{potential}}). \tag{2.4.46}$$

In the static case, the time derivative term vanishes ($E_{\text{kinematic}} = 0$), so the action is written as

$$S = -\int dt E(t)_{\text{potential}} = -\int dt M \tag{2.4.47}$$

$$\begin{aligned} M = &+ aN_c \int d^3x dz \mathrm{Tr}_f \Big[\frac{\lambda}{2} F_{MN}^2 + \Big(-\frac{z^2}{6} F_{ij}^2 + z^2 F_{iz}^2 - F_{0M}^2\Big) + \mathcal{O}(\lambda^{-1})\Big] \\ &- \frac{1}{2} aN_c \int d^3x dz \Big[\frac{\lambda}{2} \hat{F}_{MN}^2 + \Big(-\frac{z^2}{6} \hat{F}_{ij}^2 + z^2 \hat{F}_{iz}^2 - \hat{F}_{0M}^2\Big) + \mathcal{O}(\lambda^{-1})\Big] \\ &- \frac{N_c}{24\pi^2} \epsilon_{MNPQ} \int d^3x dz \Big[\frac{3}{8} \hat{A}_0 \mathrm{Tr}_f (F_{MN} F_{PQ}) - \frac{3}{2} \hat{A}_M \mathrm{Tr}_f (\partial_0 A_N F_{PQ}) \\ &+ \frac{3}{4} \hat{F}_{MN} \mathrm{Tr}_f (A_0 F_{PQ}) + \frac{1}{16} \hat{A}_0 \hat{F}_{MN} \hat{F}_{PQ} - \frac{1}{4} \hat{A}_M \hat{F}_{0N} \hat{F}_{PQ}\Big]. \end{aligned} \tag{2.4.48}$$

Substituting the classical solution, we obtain

$$\begin{aligned} M =& 8\pi^2 \kappa + \kappa \lambda^{-1} \int d^3x dz \Big[-\frac{z^2}{6} \mathrm{Tr}_f (F_{ij})^2 + z^2 \mathrm{Tr}_f (F_{iz})^2\Big] \\ &- \frac{1}{2} \kappa \lambda^{-1} \int d^3x dz \Big[(\partial_M \hat{A}_0)^2 + \frac{1}{32\pi^2 a} \hat{A}_0 \epsilon_{MNPQ} \mathrm{Tr}_f (F_{MN} F_{PQ})\Big] + \mathcal{O}(\lambda^{-1}) \\ =& 8\pi^2 \kappa \Big[1 + \lambda^{-1} \Big(\frac{\rho^2}{6} + \frac{1}{320\pi^4 a^2} \frac{1}{\rho^2} + \frac{Z^2}{3}\Big) + \mathcal{O}(\lambda^{-1})\Big] \end{aligned} \tag{2.4.49}$$

(see Sect. 2.7.4). We find that this static solution takes the minimum value

$$M_{\min} = 8\pi^2 \kappa + \sqrt{\frac{2}{15}} N_c \tag{2.4.50}$$

at

$$Z = 0 \tag{2.4.51}$$

$$\rho^2 = \frac{1}{8\pi^2 a} \sqrt{\frac{6}{5}}. \tag{2.4.52}$$

The reason for the stabilization of the solution is understood to be due to $1/\rho^2$. This $1/\rho^2$ term is originated from

$$\epsilon_{MNPQ} \int d^4x dz \hat{A}_0 \mathrm{Tr}_f (F_{MN} F_{PQ}). \tag{2.4.53}$$

The physical meaning of this term has already been explained.

In the discussion so far we have considered the static case. The baryon state could be obtained by giving the time dependence to the collective coordinate of the instanton solution and considering a quantum mechanical system in which this coordinate is the dynamical variable. We will discuss this step by step.

2.4.2 Collective Coordinates

The solution space spanned by the collective coordinates is called moduli space, which is a solution space that can move without changing its energy. By regarding these collective coordinates as dynamical variables with time dependence, a quantum mechanical system is obtained. By quantizing this system, we obtain the baryon state. As will be described later, if we include the sub-leading of the $1/\lambda$ expansion, the energy changes with the moving of (Z, ρ), and these variables are not precisely the collective coordinates. The moduli space of instanton solutions of topological number one that we are dealing with is denoted as

$$\mathcal{M} = \mathbb{R}^4 \times \mathbb{R}^4/\mathbb{Z}_2. \tag{2.4.54}$$

The first $\mathbb{R}^4$ is the space labelled by $(\vec{X}, Z)$ and is interpreted as the position of the instanton in the 4-dimensional space (x^1, x^2, x^3, z). On the other hand, $\mathbb{R}^4/\mathbb{Z}_2$ is represented by the size of the instanton ρ and its SU(2) orientation $W \in SU(2)$. This W can also be expressed as $W = \mathbf{a} = a_4 I + i a_a \tau^a$ using Pauli matrices τ^a and unit matrix I, where a_I satisfies $\sum_I a_I^2 = 1$, $(I = 1, 2, 3, 4)$. This four-dimensional manifold $\mathbb{R}^4/\mathbb{Z}_2$ can also be parameterized using the parameter y_I, which is the coordinate that labels the coset space of the four-dimensional Cartesian coordinate system by $\mathbb{Z}_2$. The y_I transforms to $y_I \to -y_I$ for $\mathbb{Z}$, where two different coordinate are related as

$$a_I = y_I/\rho \tag{2.4.55}$$

$$\rho = \sqrt{y_1^2 + y_2^2 + y_3^2 + y_4^2}. \tag{2.4.56}$$

Now, we consider the case of a slowly moving soliton. The method often used in such cases is the moduli space approximation method [15, 16]. As already mentioned several times, this method considers a quantum mechanical system in moduli space, with time dependence on the collective coordinates. It has been argued in (2.4.49) that the variables (Z, ρ) is no longer precisely the collective coordinate if we keep up to the sub-leading of $1/\lambda$. However, since the excitation energies involved in them are smaller than in the others [14], we can still deal with them as collective coordinates. This allows us to treat the baryon excitation modes.

Thus, an arbitrary solution of the gauge field with time dependence in the moduli space is denoted by

$$A_0(t,x) = \Delta A_0(t,x) \tag{2.4.57}$$

$$A_M(t,x) = W(t) A_M^{cl}(x; X^\alpha(t)) W(t)^{-1}, \tag{2.4.58}$$

with $X^\alpha = (X^M, \rho)$. We use the notation $x^0 = t$ since we consider time and space independently. So far, there seems to be no reason why the time component of the gauge field should be induced in (2.4.57). The induced time component is an essential requirement in the gauge theory (Appendix 2.6). We can also understand this time component from the fact that the EOM (2.4.31) would not be satisfied without this induced term.

If we give a time dependence to the collective coordinates of the solution, this solution only moves in moduli space and is still expected to be a solution of (2.4.32). However, this does not mean that the other equations of motion are unchanged. Therefore, we will check whether the solution given the time dependence satisfies the equations of motion again.

First, after giving a time dependence to the gauge field in collective coordinates, the field strength is written as

$$F_{MN} = W F_{MN}^{cl} W^{-1} \tag{2.4.59}$$

$$\begin{aligned} F_{0M} &= \partial_0 A_M - \partial_M \Delta A_0 + i[\Delta A_0, A_M] \\ &= W\left(\dot{X}^\alpha \frac{\partial}{\partial X^\alpha} A_M^{cl} - D_M^{cl} \Phi\right). \end{aligned} \tag{2.4.60}$$

The calculation of F_{0M} is shown in Appendix 2.7.5, where Φ is defined by

$$\Phi = W^{-1} \Delta A_0 W - W^{-1} \dot{W}. \tag{2.4.61}$$

With this Φ, (2.4.57) and (2.4.58) are together denoted as

$$\mathcal{A}(t,x) = \left(\mathcal{A}^{cl}(x; X^\alpha(t)) + \Phi(t,x) dt\right)^{W(t)}. \tag{2.4.62}$$

The calligraphic letter in the notation when including to SU(2) and U(1) terms. In the above equation, a 5-form of (t, x^1, x^2, x^3, z), which is a shorthand for $A^{V(t,x)} = V^{-1} A V + i V^{-1} d V$.

From the above, (2.4.32) becomes

$$D_N F_{MN} = 0 \rightarrow D_N^{cl} F_{MN}^{cl} = 0, \tag{2.4.63}$$

which does not change the form of the equation of motion when the solution has a time dependence. Therefore, the solution is still BPST instanton;

$$A_M(x) = -i f(\xi(t)) g(x; X^\alpha(t)) \partial_M g(x; X^\alpha(t))^{-1}. \tag{2.4.64}$$

This is natural since it is just the definition of the collective coordinates. We also see that (2.4.34) does not change its form, hence (2.4.41) does not change either, indicating that we can still use

$$\hat{A}_0 = \frac{1}{8\pi^2 a}\frac{1}{\xi(t)^2}\Big(1 - \frac{\rho(t)^4}{(\xi(t)^2+\rho(t)^2)^2}\Big) \tag{2.4.65}$$

$$\hat{A}_M = 0 \tag{2.4.66}$$

as a solution to the equation of motion. So, the three solutions only change the collective coordinate from constants to variables with time dependence.

The only equation that changes is (2.4.31), which becomes

$$D^{cl}_M\Big(\dot{X}^\alpha \frac{\partial}{\partial X^\alpha} A^{cl}_M - D^{cl}_M \Phi\Big) = 0, \tag{2.4.67}$$

where we use the Eq. (2.4.60), The Φ is determined by solving this equation. As shown in Appendix 2.7.6, the solution of $\Phi(t, x)$ is obtained as follow;

$$\Phi(t,x) = -\dot{X}^N(t)A^{cl}_N(x) + \chi^a(t)\Phi_a(x) \tag{2.4.68}$$

$$\Phi_a = f(\xi)g(x; X^\alpha(t))t_a g(x; X^\alpha(t))^{-1} \tag{2.4.69}$$

$$\chi^a(t) = -2i\,\mathrm{tr}(t_a W^{-1}\dot{W}) = 2(a_4\dot{a}_a - \dot{a}_4 a_a + \epsilon_{abc}a_b\dot{a}_c). \tag{2.4.70}$$

Here, it is useful to have the following equation;

$$g\partial_M g^{-1} = \begin{cases} \frac{i}{\xi^2}\big((z-Z)\tau^i - \epsilon_{ija}(x^j - X^j)\tau^a\big) \\ -\frac{i}{\xi^2}(x^a - X^a\tau^a) \end{cases} \tag{2.4.71}$$

$$\partial_M(g\partial_M g^{-1}) \propto (x^M - X^M)g\partial_M g^{-1} = 0. \tag{2.4.72}$$

and spin and isospin are defined by

$$J_i = 8\pi^2\kappa\rho^2\mathrm{tr}(-iW^{-1}\dot{W}t_i), \tag{2.4.73}$$

$$I_a = 8\pi^2\kappa\rho^2\mathrm{tr}(i\dot{W}W^{-1}t_a) = -WJ_it_iW^{-1}. \tag{2.4.74}$$

By introducing the time dependence, a kinetic term of (2.4.29) appears. We can now identify the coefficients (mass) of the kinetic term with the metrics of the moduli space (the kinetic term in the U(1) term is still zero) which is written as

$$+aN_c\int d^4x dz \mathrm{Tr}_f F^2_{0M} = aN_c\int d^4x dz \mathrm{Tr}_f(D^{cl}_M\Phi - \dot{A}^{cl}_M)^2$$
$$= \frac{m_{X^\alpha}}{2} g_{\alpha\beta}\dot{X}^\alpha\dot{X}^\beta. \tag{2.4.75}$$

Since $X^{\alpha} = (\mathbf{X}, X^Z, \rho)$ are found to be orthogonal to each other, the mass is written as $m_{X^{\alpha}}$. From the above, Lagrangian is obtained as

$$S = \int dt L = \int dt (L_{\text{kinetic}} - L_{\text{potential}}) \tag{2.4.76}$$

$$L = L_X + L_Z + L_y + \mathcal{O}(\lambda^{-1}) \tag{2.4.77}$$

$$L_X = -M_0 + \frac{m_X}{2}\dot{\mathbf{X}}^2 \tag{2.4.78}$$

$$L_Z = \frac{m_Z}{2}\dot{Z}^2 - \frac{m_Z \omega_Z^2}{2} Z^2 \tag{2.4.79}$$

$$L_y = \frac{m_y}{2}(\dot{\rho}^2 + \rho^2 \dot{a}_I^2) - \frac{m_y \omega_\rho^2}{2}\rho^2 - \frac{Q}{\rho^2} \tag{2.4.80}$$

$$= \frac{m_y}{2}\dot{y}_I^2 - \frac{m_y \omega_\rho^2}{2} y_I^2 - \frac{Q}{\rho^2}. \tag{2.4.81}$$

The plus term in L is the result of the time dependence, and the minus term is the original potential term (2.4.49), with

$$M_0 = 8\pi^2\kappa, \quad m_X = m_Z = \frac{m_y}{2} = 8\pi^2\kappa\lambda^{-1} = 8\pi^2 a N_c$$
$$\omega_Z^2 = \frac{2}{3}, \quad \omega_\rho^2 = \frac{1}{6}, \quad Q = \frac{N_c^2}{5 m_X} = \frac{N_c}{40\pi^2 a}. \tag{2.4.82}$$

As described above, we have obtained a quantum mechanical system with a finite number of time-dependent collective coordinates as dynamical variables.

2.4.3 Collective Coordinates Quantization

The derived mechanical system is quantized to obtain the spectra of baryons. We consider here the case where the position of the instanton, the baryon, is $\vec{X} = 0$. The collective coordinate quantization described here is often used in the quantization of solitons [18, 21–23].

For the dynamical variable (Z, ρ, a_I) in the Lagrangian, we impose the canonical commutation relation and quantize the system. Then, we obtain the Hamiltonian,

$$H = M_0 + H_y + H_Z \tag{2.4.83}$$

$$H_y = -\frac{1}{2m_y}\Big(\frac{1}{\rho^3}\partial_\rho(\rho^3\partial_\rho) + \frac{1}{\rho^2}(\nabla_{S^3}^2 - 2m_y Q)\Big) + \frac{1}{2} m_y \omega_{\rho^2} \rho^2 \tag{2.4.84}$$

$$H_Z = -\frac{1}{2m_Z}\partial_Z^2 + \frac{1}{2} m_Z \omega_Z^2 Z^2, \tag{2.4.85}$$

where $\nabla^2_{S^3}$ is Laplacian on the S^3.

Since the H_Z is the Hamiltonian of the harmonic oscillator, its eigenvalue is

$$E_Z = \omega_Z\left(n_Z + \frac{1}{2}\right) = \frac{2n_Z + 1}{\sqrt{6}}. \tag{2.4.86}$$

The eigenvalues and eigenfunctions of the Laplacian are given by

$$\nabla^2_{S^3} T^{(l)} = -l(l+2)T^{(l)}. \tag{2.4.87}$$

Here, $T^{(l)}$ is the spherical harmonics on S^3, which is expressed as

$$T^{(l)} = C_{I_1\cdots I_l} a_{I_1} \cdots a_{I_l} \tag{2.4.88}$$

with the traceless symmetric tensor $C_{I_1\cdots I_l}$ of rank l. Using this function, we write the eigenfunctions of the H_y in the variable-separated form

$$\psi(y_I) = R_l(\rho)T^{(l)}(a_I). \tag{2.4.89}$$

When operate H_y on $R_l(\rho)$, we obtain the eigenvalue equations,

$$\mathcal{H}_l R_l(\rho) = E_y R_l(\rho) \tag{2.4.90}$$

$$\mathcal{H}_l = -\frac{1}{2m_y}\left(\frac{1}{\rho^3}\partial_\rho(\rho^3\partial_\rho) - \frac{l(l+2) + 2m_y Q}{\rho^2}\right) + \frac{1}{2}m_y\omega_\rho^2\rho^2. \tag{2.4.91}$$

Here, if we replace

$$\tilde{l} = -1 + \sqrt{(l+1)^2 + 2m_y Q}, \quad \tilde{l}(\tilde{l}+2) = l(l+2) + 2m_y Q, \tag{2.4.92}$$

we can write it as

$$\mathcal{H}_l = -\frac{1}{2m_y}\left(\frac{1}{\rho^3}\partial_\rho(\rho^3\partial_\rho) - \frac{\tilde{l}(\tilde{l}+2)}{\rho^2}\right) + \frac{1}{2}m_y\omega_\rho^2\rho^2. \tag{2.4.93}$$

Now, if we write

$$R_l(\rho) = \exp\left(-\frac{m_y\omega_\rho}{2}\rho^2\right)\rho^{\tilde{l}} v(m_y\omega_\rho\rho^2), \tag{2.4.94}$$

with $x = m_y\omega_\rho\rho^2$, we see that $v(x)$ must be satisfied by

$$\left(x\partial_x + (\tilde{l} + 2 - x)\partial_x + \frac{1}{2}\left(\frac{E_y}{\omega\rho} - \tilde{l} - 2\right)\right)v(x) = 0. \tag{2.4.95}$$

This is a confluent hypergeometric differential equation for $v(x)$, and a normalizable and regular solution exists only at

$$\frac{1}{2}\left(\frac{E_y}{\omega\rho} - \tilde{l} - 2\right) = n \in \mathbb{Z}. \tag{2.4.96}$$

Since $v(x = m_y\omega_\rho\rho^2)$ can be labeled by n and $\tilde{l}$, if we denote $v(x = m_y\omega_\rho\rho^2) = F(-n, \tilde{l}; x)$, then the confluent hypergeometric function

$$F(\alpha, \gamma; z) = \sum_{k=0}^{\infty} \frac{(\alpha)_k}{(\gamma)_k}\frac{z^k}{k!} \tag{2.4.97}$$

$$(\alpha)_k = \alpha(\alpha+1)\cdots(\alpha+k-1) \tag{2.4.98}$$

satisfies

$$\left(x\partial_x + (\tilde{l} + 2 - x)\partial_x + n\right)F(-n, \tilde{l}; x) = 0. \tag{2.4.99}$$

Using $F(-n, \tilde{l}; m_y\omega_\rho\rho^2)$, we can write

$$R(\rho) = \exp\left(-\frac{m_y\omega_\rho}{2}\rho^2\right)\rho^{\tilde{l}}F(-n, \tilde{l}; m_y\omega_\rho\rho^2), \tag{2.4.100}$$

so the eigenvalue is obtained as

$$E_y = \omega_\rho(\tilde{l} + 2n + 2) \tag{2.4.101}$$

$$= \sqrt{\frac{(l+1)^2}{6} + \frac{2}{15}N_c^2} + \frac{2n_\rho + 1}{\sqrt{6}}. \tag{2.4.102}$$

From the above, the eigenvalues of H are

$$M = M_0 + E_y + E_Z \tag{2.4.103}$$

$$= M_0 + \sqrt{\frac{(l+1)^2}{6} + \frac{2}{15}N_c^2} + \frac{2(n_\rho + n_Z + 1)}{\sqrt{6}} \tag{2.4.104}$$

which is the baryon mass formula we desired.

2.4.4 Physical Interpretation of Quantum Numbers

We discuss the physical interpretation of quantum numbers appearing in the mass formula. Considering the $SU(2)_L \times SU(2)_R$ transformation for $W(t)$;

Table 2.3 The spectra of $N(l = 1)$

(n_ρ, n_z)	(0, 0)	(1, 0)	(0, 1)	(1, 1)	(2, 0)/(0, 2)	(2, 1)/(0, 3)	(1, 2)/(3, 0)
Prediction	[940$^+$]	1348$^+$	1348$^-$	1756$^-$	1756$^+$, 1756$^-$	2164$^-$, 2164$^+$	2164$^+$, 2164$^+$
Experiment	940$^+$	1440$^+$	1535$^-$	1655$^-$	1710$^+$, ?	2090$^-$, ?	2100$^+$, ?

Table 2.4 The spectra of $\Delta(l = 3)$

(n_ρ, n_z)	(0, 0)	(1, 0)	(0, 1)	(1, 1)	(2, 0)/(0, 2)	(2, 1)/(0, 3)	(1, 2)/(3, 0)
Prediction	1240$^+$	1648$^+$	1648$^-$	2056$^-$	2056$^+$, 2056$^-$	2464$^-$, 2464$^+$	2464$^+$, 2464$^+$
Experiment	1232$^+$	1600$^+$	1700$^-$	1940$^-$	1920$^+$, ?	?, ?	?, ?

$$W(t) \to g_L W(t) g_R, \quad g_{L/R} \in SU(2)_{L/R} \tag{2.4.105}$$

the gauge field A_M transforms as

$$\begin{aligned} A_M(t, x^N) &= W(t) A_M^{cl}(x^N) W(t)^{-1} \\ \to & g_L W(t) g_R A_M^{cl}(x^N) g_R^{-1} W(t)^{-1} g_L^{-1} \\ &= g_L A_M(t, Rx^N) g_L^{-1}. \end{aligned} \tag{2.4.106}$$

This means that g_L corresponds to the isospin I_a and g_R, which causes spatial rotation, corresponds to the spin J_a. Furthermore, as discussed in the Sect. 3.1.3, we see that the relation

$$\text{tr}(I_a^2) = \text{tr}(J_a^2) \tag{2.4.107}$$

$$I = J = \frac{l}{2}. \tag{2.4.108}$$

Let us now look at the quantum number (n_ρ, n_Z). The parity transformation is defined by $P : (t, x^M) \to (t, -x^M)$, as discussed in the Sect. 2.3.3. The baryons with odd (even) quantum numbers n_Z correspond to negative (positive) parity excitation, because the wave functions are odd (even) with respect to the $Z \to -Z$ transformation. The states of each baryon obtained from the Sakai-Sugimoto model should be identified with the states shown in Tables 2.3 and 2.4 with $M_{KK} = 500$.

Finally, we comment on the obtained baryon mass formula. At large N_c, the mass formula (2.4.103) is approximately written as

$$N_c \gg 1 \tag{2.4.109}$$

$$M \simeq M_0 + \sqrt{\frac{2}{15}} N_c + \frac{1}{4}\sqrt{\frac{5}{6}} \frac{(l+1)^2}{N_c} + \frac{2(n_\rho + n_Z + 1)}{\sqrt{6}}. \tag{2.4.110}$$

Considering up to the order of $\mathcal{O}(N_c)$, we see that it is a static baryon mass formula (2.4.50). Interestingly, the N_c dependence of (2.4.109) has the same form as the result of Refs. [18, 24] in large N_c QCD. This means that the mass splitting between states of the same internal excitation states of different spins is of the order of $1/N_c$, and the mass splitting between the internal excitation states of each other is of the order of N_c^0.

The Wave Functions of Baryon States

Here is a summary of the baryon wave function. The canonical momentum of $(\vec{X}, Z, y_I)$ is

$$\vec{P} = M_0\dot{\vec{X}} = -i\frac{\partial}{\partial \vec{X}}, \quad P_Z = M_0\dot{Z} = -i\frac{\partial}{\partial Z}, \quad \Pi_I = 2M_0\dot{y}_I = -i\frac{\partial}{\partial y_I}. \tag{2.4.111}$$

Also, as mentioned above, the left rotation g_L of W corresponds to isospin I_a and the right rotation g_R of W corresponds to spin J_a, so isospin and spin are given as

$$I_a = \frac{i}{2}\left(y_4\frac{\partial}{\partial y_a} - y_a\frac{\partial}{\partial y_4} - \epsilon_{abc}y_b\frac{\partial}{\partial y_c}\right) \tag{2.4.112}$$

$$J_a = \frac{i}{2}\left(-y_4\frac{\partial}{\partial y_a} + y_a\frac{\partial}{\partial y_4} - \epsilon_{abc}y_b\frac{\partial}{\partial y_c}\right). \tag{2.4.113}$$

See also the discussion of Appendix 4.4 for this. The Hamiltonian is given by

$$H = \frac{1}{2M_0}(\vec{P}^2 + P_Z^2) + \frac{1}{4M_0}\Pi_I^2 + U(\rho, Z) \tag{2.4.114}$$

$$U(\rho, Z) = 8\pi^2\kappa\left(1 + \frac{\rho^2}{6} + \frac{N_c^2}{320\pi^4\kappa^2}\frac{1}{\rho^2} + \frac{Z^2}{3}t\right). \tag{2.4.115}$$

The eigenfunctions of this Hamiltonian are characterized by $(l, I_3, J_3, n_\rho, n_z)$ and momentum $\vec{p}$ and are denoted by

$$|\vec{p}, B, s\rangle = |\vec{p}\rangle\,|B\rangle\,, \quad |\vec{p}\rangle = \frac{1}{(2\pi)^{3/2}}e^{i\vec{p}\cdot\vec{X}}, \tag{2.4.116}$$

$$|B\rangle = |l, I_3, J_3\rangle\,|n_\rho\rangle\,|n_Z\rangle\,. \tag{2.4.117}$$

In the following, we omit the normalization constants and provide specific wave functions. First, with respect to $|l, I_3, J_3\rangle$, for example, it is given as

$$|l = 1, I_3 = J_3 = l/2\rangle = a_1 + ia_2, \tag{2.4.118}$$

$$|l = 3, I_3 = J_3 = l/2\rangle = (a_1 + ia_2)^3. \tag{2.4.119}$$

From this expression, using the ladder operator $I_- = I_1 - iI_2$, we get

$$|l = 1, I_3 = -l/2, J_3 = l/2\rangle = I_-(a_1 + ia_2) = a_4 + ia_3. \tag{2.4.120}$$

Also, with respect to $|\rho\rangle$, we obtain

$$|n_\rho = 0\rangle = \rho^{-1+2\sqrt{1+N_c^2/5}} e^{-\frac{M_0}{\sqrt{6}}\rho^2}, \tag{2.4.121}$$

$$|n_\rho = 1\rangle = \left(\frac{2M_0}{\sqrt{6}}\rho^2 - 1 - 2\sqrt{1+N_c^2/5}\right)\rho^{-1+2\sqrt{1+N_c^2/5}} e^{-\frac{M_0}{\sqrt{6}}\rho^2}. \tag{2.4.122}$$

Finally, for $|Z\rangle$, we have

$$|n_Z = 0\rangle = e^{-\frac{M_0}{\sqrt{6}}Z^2}, \tag{2.4.123}$$

$$|n_Z = 1\rangle = Ze^{-\frac{M_0}{\sqrt{6}}Z^2}. \tag{2.4.124}$$

From the above, the wave functions of the proton and neutron are

$$|p\rangle = \rho^{-1+2\sqrt{1+N_c^2/5}} e^{-\frac{M_0}{\sqrt{6}}\rho^2} e^{-\frac{M_0}{\sqrt{6}}Z^2}(a_1 + ia_2) \tag{2.4.125}$$

$$|n\rangle = \rho^{-1+2\sqrt{1+N_c^2/5}} e^{-\frac{M_0}{\sqrt{6}}\rho^2} e^{-\frac{M_0}{\sqrt{6}}Z^2}(a_4 + ia_3), \tag{2.4.126}$$

respectively (see also Appendix 4.4). Also, $\Delta(1232)^{++}$ is represented as

$$\left|\Delta(1232)^{++}\right\rangle = \rho^{-1+2\sqrt{1+N_c^2/5}} e^{-\frac{M_0}{\sqrt{6}}\rho^2} e^{-\frac{M_0}{\sqrt{6}}Z^2}(a_1 + ia_2)^3. \tag{2.4.127}$$

Furthermore, the wave functions of the Roper and negative parity resonances are given by

$$\left|N^*(1440)\right\rangle = \left(\frac{2M_0}{\sqrt{6}}\rho^2 - 1 - 2\sqrt{1+N_c^2/5}\right)\rho^{-1+2\sqrt{1+N_c^2/5}} e^{-\frac{M_0}{\sqrt{6}}\rho^2} \times e^{-\frac{M_0}{\sqrt{6}}Z^2}(a_1 + ia_2) \tag{2.4.128}$$

$$\left|N^*(1535)\right\rangle = \rho^{-1+2\sqrt{1+N_c^2/5}} e^{-\frac{M_0}{\sqrt{6}}\rho^2} Ze^{-\frac{M_0}{\sqrt{6}}Z^2}(a_1 + ia_2). \tag{2.4.129}$$

2.5 Appendix: Dp-Brane and Black p-Brane

In this Appendix, we will briefly discuss topics related to superstring theory and supergravity theory that are omitted in the text for convenience of explanation, based on Ref. [25].

2.5.1 Low-Energy Effective Action of D-Branes

The coordinates of the worldsheet of Dp-brane are ξ^a with $a = 0, 1, \ldots, p$. The gauge field part of the Dp-brane action (DBI action) with constant dilaton $e^{\phi_0} = g_s$ is given by

$$\begin{aligned} S_{\rm DBI} &= -T_p \int d^{p+1}\xi e^{-\phi_0}\sqrt{-\det(P[g]_{ab} + 2\pi\alpha' F_{ab})} \\ &= -\frac{T_p}{g_s}\int d^{p+1}\xi\sqrt{-\det(P[g]_{ab} + 2\pi\alpha' F_{ab})}, \end{aligned} \tag{2.5.1}$$

where $P[g]$ are the pullback for bulk metric g_{MN} $(M, N = 0, 1, \ldots 9)$ as follow,

$$P[g]_{ab} = \frac{\partial X^M}{\partial \xi^a}\frac{\partial X^N}{\partial \xi^b} g_{MN}, \tag{2.5.2}$$

with tension for Dp-brane T_p. Considering the embedding of this action in flat space, if we expand using

$$\det(1 + M) = 1 - \frac{1}{2}\mathrm{Tr}(M^2) + \cdots, \tag{2.5.3}$$

we obtain

$$S_{\rm DBI} = -(2\pi\alpha')^2 \frac{T_p}{4g_s}\int d^{p+1}\xi F_{ab}F^{ab} \tag{2.5.4}$$

from the lowest order of α'. This implies that the DBI action for a single Dp-brane is a generalization of the U(1) Yang-Mills theory. Thus, from this action, we find a correspondence,

$$g_{YM}^2 = \frac{g_s}{T_p(2\pi\alpha')^2} = (2\pi)^{p-2} g_s \alpha'^{\frac{p-3}{2}}. \tag{2.5.5}$$

2.5.2 Black p-Brane Solution

The low energy effective action of Type IIB superstring theory is reduced to a supergravity action. The bosonic part of Type IIB supergravity action is written as

$$\begin{aligned} S_{\rm IIB} = \frac{1}{2\tilde{\kappa}_{10}^2}\Big[\int d^{10}X\sqrt{-g}\Big(e^{-2\phi}\Big(R + 4\partial_M\phi\partial^M\phi - \frac{1}{2}|H_{(3)}|^2\Big) - \frac{1}{2}|F_{(1)}|^2 \\ -\frac{1}{2}|\tilde{F}_{(3)}|^2 - \frac{1}{4}|\tilde{F}_{(5)}|^2\Big) - \frac{1}{2}\int C_{(4)} \wedge H_{(3)} \wedge F_{(3)}\Big] \end{aligned} \tag{2.5.6}$$

where $\bar{F}_{(p)}$ is the conplex conjugate of $F_{(p)}$, using the notation of

$$\int d^{10}X\sqrt{-g}|F_{(p)}|^2 = \frac{1}{p!}\int d^{10}X\sqrt{-g}g_{M_1N_1}\cdots g_{M_pN_p}\bar{F}^{M_1\cdots M_p}F^{N_1\cdots N_p}. \quad (2.5.7)$$

A comparison with superstring theory shows that the ten-dimensional gravitational constant $\tilde{\kappa}_{10}$ is given by

$$2\tilde{\kappa}_{10}^2 = (2\pi)^7\alpha'^4. \quad (2.5.8)$$

Considering the asymptotic value of dilaton and using

$$2\kappa_{10}^2 = 2\tilde{\kappa}_{10}^2 g_s^2 = (2\pi)^7\alpha'^4 g_s^2, \quad (2.5.9)$$

the ten-dimensional Newton constant G_{10} is defined as $G_{10} = \kappa_{10}^2/8\pi$. The field strengths are defined by

$$F_{(p)} = \mathrm{d}C_{(p-1)},\ H_{(3)} = \mathrm{d}B_{(2)},\ \tilde{F}_{(3)} = F_{(3)} - C_{(0)}H_{(3)}, \quad (2.5.10)$$

$$\tilde{F}_{(5)} = F_{(5)} - \frac{1}{2}C_{(2)}\wedge H_{(3)} + \frac{1}{2}B_{(2)}\wedge F_{(3)} \quad (2.5.11)$$

respectively, and d denotes the exterior derivative.

The ten-dimensional Dp-brane has symmetry of $\mathbb{R}^{p+1}\times SO(p,1)\times SO(9-p)$. The ansatz for solving the equation of motion of the type IIB supergravity is

$$ds^2 = H_p(r)^{-1/2}\eta_{\mu\nu}dx^\mu dx^\nu + H_p(r)^{1/2}dy^i dy^i, \quad (2.5.12)$$

$$e^\phi = g_s H_p(r)^{(3-p)/4}, \quad (2.5.13)$$

$$C_{(p+1)} = \big(H_p(r)^{-1} - 1\big)dx^0\wedge dx^1\wedge\cdots\wedge dx^p, \quad (2.5.14)$$

$$B_{MN} = 0, \quad (2.5.15)$$

where x^μ $(\mu = 0,\ldots,p)$ are the coordinates on the brane worldvolume, and y^i $(i = p+1,\ldots,9)$ describe the coordinates perpendicular to the brane with $r^2 = \sum_{i=p+1}^9 y_i^2$. Substituting this ansatz into the equation of motion of type IIB supergravity (2.5.6) leads to

$$\Box H_p(r) = 0 \quad (2.5.16)$$

at $r\neq 0$, where $\Box = \partial_i\partial^i$. Therefore, we can determine

$$H_p(r) = 1 + \left(\frac{r_0^p}{r}\right)^{7-p} \quad (2.5.17)$$

using the asymptotic to the ten-dimensional Minkowski spacetime at $r\to\infty$ away from the brane. This charge can be determined by calculating the integral

$$Q = \frac{1}{2\kappa_{10}^2}\int_{S^{8-p}} *F_{(p+2)} \tag{2.5.18}$$

of the R-R flux on the $(8-p)$-dimensional sphere at infinity of the $(9-p)$-dimensional transversal space of the Dp-brane, where * means the ten-dimensional Hoge dual operator. If the charge of one Dp-brane is 1, the total charge of N Dp-branes is N. From the above, it is determined to be

$$(r_0^p)^{7-p} = (4\pi)^{(5-p)/2}\Gamma\left(\frac{7-p}{2}\right) g_s N \alpha'^{(7-p)/2}. \tag{2.5.19}$$

2.5.3 *Black p-Brane at Near Horizon*

In the limit $r \to 0$ with $u = r/l_s^2 =$ fixed, H in Eq. (2.5.17) reduces to

$$H = 1 + \frac{4\pi g_s N_c l_s^4}{r^4} = 1 + \frac{2g_{YM}^2 N_c l_s^4}{u^4 l_s^8} \to \frac{2g_{YM}^2 N_c}{u^4 l_s^4}, \tag{2.5.20}$$

where 't Hooft coupling $\lambda = g_{YM}^2 N_c$, the effective coupling constant in gauge theory, also kept finite. From this, using u, the black 3-brane spacetime is written as

$$\frac{ds^2}{l_s^2} = \frac{u^2}{u_0^2}(-dt^2 + d\vec{x}^2) + \frac{u_0^2}{u^2}du^2 + u_0^2 d\Omega_5^2, \tag{2.5.21}$$

where $u_0^4 = (r_0/l_s)^4 = 2g_{YM}^2 N_c$. This is the equation in the near horizon of the black 3-brane. Thus, in the near horizon, it is written as a direct product of AdS_5 (first two terms) and S^5 $(u_0^2 d\Omega_5^2)$.

2.5.4 *Conical Singularity*

Consider the case where the metric has the form

$$ds^2 = \cdots + f(r)dt^2 + \frac{dr^2}{f(r)} + \cdots \tag{2.5.22}$$

with $f(r \sim r_0) = 0$. This form clearly appears to have a singularity. Now, by Taylor expansion around $r \sim r_0$, $f \sim f'(r_0)(r - r_0)$, metric can be written as

$$ds^2 = \cdots + f'(r_0)(r - r_0)dt^2 + \frac{dr^2}{f'(r_0)(r - r_0)} + \cdots \tag{2.5.23}$$

If we transform the coordinates to

$$\rho = 2\sqrt{\frac{(r - r_0)}{f'(r_0)}}, \tag{2.5.24}$$

we can write metric as

$$ds^2 = \cdots + d\rho^2 + \rho^2 d\Big(\frac{f'(r_0)}{2}t\Big)^2 + \cdots \tag{2.5.25}$$

Then, if $\frac{f'(r_0)}{2}t$ has periodically 2π, this metric has the exact same form as the polar coordinates. Therefore, in order not to have singularity, t must have

$$\delta t = \frac{f'(r_0)}{4\pi} \tag{2.5.26}$$

periodicity.

2.6 Appendix: Gauss's Law

In this doctoral thesis, I mentioned that one should pay attention to whether Gauss's law is satisfied when we give time dependence to the static soliton solutions in gauge theory by the collective coordinate. Gauss' law is an equation that determines the time component of the gauge field, which is not an equation of time evolution because it does not include the time derivative for the time component of the gauge field (due to the fact that the field strength is completely antisymmetric). In fact, this time component is not an independent dynamical variable that disappears in the Hamiltonian. Let us now follow the textbook [26] to clarify what is implied by Gauss's law and to better understand the collective coordinate quantization of solitons in gauge theory.

In the following, for simplicity, let us consider a U(1) gauge theory of

$$\mathcal{L} = -\frac{1}{4} f_{\mu\nu} f^{\mu\nu} + \frac{1}{2}\overline{D_\mu\phi} D^\mu\phi \tag{2.6.1}$$

with gauge fields a_μ, $(\mu = 0, 1, 2, 3)$ and scalar fields ϕ, where $f_{\mu\nu}$ is field strength and $D_\mu\phi = \partial_\mu\phi + i a_\mu\phi$ covariant derivative ($\overline{D_\mu\phi}$ is its complex conjugate.). Gauss' law is the time component of the EOM of a_μ which is obtained as

$$(\nabla^2 - \overline{\phi}\phi)a_0 = \partial_i\partial_0 a_i + \frac{i}{2}\big(\overline{\phi}\partial_0\phi - \phi\partial_0\overline{\phi}\big). \tag{2.6.2}$$

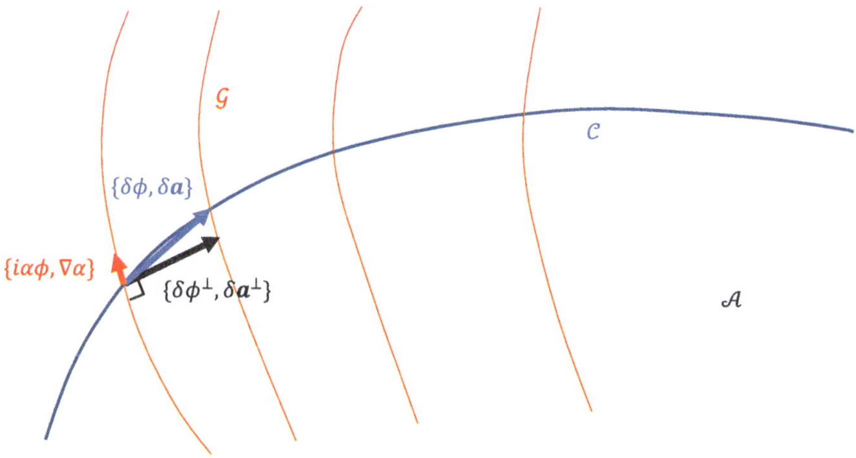

Fig. 2.5 The coordinate space of the gauge and scalar fields

This equation implies that a_0 is determined if the time derivatives of a_i, $(i = 1, 2, 3)$ and ϕ at a certain time are known, which means that a_0 is an independent dynamical variable.

To understand the role of a_0, we consider the coordination space $\mathcal{A}$, which is the set of fields $\{\phi(\mathbf{x}), \mathbf{a}(\mathbf{x})\}$ at a certain time (Fig. 2.5). The physical configuration space is the coset space $\mathcal{C} = \mathcal{A}/\mathcal{G}$, which is divided by the configuration space shifted by the gauge transformation $\mathcal{G} = \{e^{i\alpha(\mathbf{x})}\}$.

Suppose that the coordination of a field, which was $\{\phi, \mathbf{a}\}$ at a certain time t, changes to $\{\phi + \delta\phi, \mathbf{a} + \delta\mathbf{a}\}$ with a infinitesimal shift $t + \delta t$. Then, the kinetic energy changes to

$$\frac{1}{2}\int \frac{1}{(\delta t)^2}\left(\delta\mathbf{a}\cdot\delta\mathbf{a} + \overline{\delta\phi}\delta\phi\right)d^4x. \tag{2.6.3}$$

If there is a component along the direction $\{i\alpha\phi, \nabla\alpha\}$ of $\mathcal{G}$ at t due to (2.6.3), the energy will be changed by the gauge transformation, which means that the gauge symmetry of the theory will be broken. To avoid such a situation, we should restrict the time variation of the field configuration to $\{\delta\phi^{\perp}, \delta\mathbf{a}^{\perp}\}$ of

$$\delta\phi^{\perp} = \delta\phi - i\beta\phi \tag{2.6.4}$$

$$\delta\mathbf{a}^{\perp} = \delta\mathbf{a} - \nabla\beta. \tag{2.6.5}$$

The β is determined from the condition that $\{\delta\phi^{\perp}, \delta\mathbf{a}^{\perp}\}$ and $\{i\alpha\phi, \nabla\alpha\}$ are orthogonal,

$$\int\left(\delta\mathbf{a}^{\perp}\cdot\nabla\alpha + \frac{1}{2}\left(\delta\phi^{\perp}\overline{(i\alpha\phi)} + \overline{(\delta^{\perp})}i\alpha\phi\right)\right)d^4x = 0. \tag{2.6.6}$$

In this way, it is understood that the time component of the gauge field a_0 plays the role of β, which limits arbitrary time changes of $\{\phi(\mathbf{x}), \mathbf{a}(\mathbf{x})\}$ at a certain time so that they are physically consistent, as follows.

Integrating (2.6.6) by parts and dropping the surface term, we get

$$\int \left(\nabla \cdot \delta \mathbf{a}^{\perp} + \frac{i}{2}\left(\overline{\phi}\delta\phi^{\perp} - \phi\overline{\delta\phi^{\perp}}\right)\right)\alpha d^4x = 0. \tag{2.6.7}$$

Since this should hold for any infinitesimal α, it is required that

$$\nabla \cdot \delta \mathbf{a}^{\perp} + \frac{i}{2}\left(\overline{\phi}\delta\phi^{\perp} - \phi\overline{\delta\phi^{\perp}}\right) = 0. \tag{2.6.8}$$

Substituting (2.6.4) and (2.6.5), we obtain

$$\left(\nabla^2 - \overline{\phi}\phi\right)\beta = \nabla \cdot \delta \mathbf{a} + \frac{i}{2}\left(\overline{\phi}\delta\phi^{\perp} - \phi\overline{\delta\phi^{\perp}}\right). \tag{2.6.9}$$

Dividing both sides by δt, and from $\delta\mathbf{a}/\delta t = \partial_0\mathbf{a}$, $\delta\phi/\delta t = \partial_0\phi$, we obtain Gauss's law (2.6.2) by regarding $\beta/\delta t = a_0$.

Thus, even if the time component of the gauge field is zero to begin with, the time component is induced when the gauge field at a certain time t is time development. This β corresponds to $\Delta A_0(t, x)$ in (2.4.57) when the gauge field is given a time dependence by the collective coordinate.

2.7 Appendix: Details of the Calculation

2.7.1 Equation (2.4.19)

We show the following expression for the CS term;

$$\begin{aligned} S_{CS} &= \frac{N_c}{24\pi^2}\int_{\mathbb{R}^4\times[0,\infty)} \mathrm{Tr}_f\left(\mathcal{A}\mathcal{F}^2 - \frac{i}{2}\mathcal{A}^3\mathcal{F} - \frac{1}{10}\mathcal{A}^5\right) \\ &= \frac{N_c}{24\pi^2}\int \Big[\omega_5(A) + \frac{3}{\sqrt{2N_f}}\hat{A}\mathrm{Tr}_f F^2 + \frac{1}{2\sqrt{2N_f}}\hat{A}\hat{F}^2 \\ &+ \frac{1}{\sqrt{2N_f}}\mathrm{d}\left(\hat{A}\mathrm{Tr}_f\left(2FA - \frac{i}{2}A^3\right)\right)\Big]. \end{aligned} \tag{2.7.1}$$

All products are wedge products.

1. $\hat{F} = d\hat{A}$, $F = dA + iA^2$, $\mathrm{Tr}_f(T^aT^b) = \frac{1}{2}\delta^{ab}$.
2. From the properties of wedge product, $\hat{A}A + A\hat{A} = \hat{A}A - \hat{A}A = 0$.
3. From the antisymmetric properties of wedge product, $\hat{A}^2 = 0$.

4. The terms of one power of the field (F and A) in $SU(N_f)$ are zero from traceless.
5. Since the coupling constant is absorbed in the gauge field, more than three squares of the $SU(N_f)$ field is zero.
6. From the properties of trace and antisymmetry of the wedge product, $\mathrm{Tr}_f A^2 = 0$.

are used as needed, where

$$\mathcal{A} = A^a T^a + \frac{1}{\sqrt{2N_f}}\hat{A} I. \tag{2.7.2}$$

Substituting

$$\mathcal{A}^2 = \left(A + \frac{1}{\sqrt{2N_f}}\hat{A}\right)^2 = A^2 + \frac{1}{\sqrt{2N_f}}(A\hat{A} + \hat{A}A) + \frac{1}{2N_f}\hat{A}^2 = A^2 \tag{2.7.3}$$

$$\mathcal{A}^3 = \left(A + \frac{1}{\sqrt{2N_f}}\hat{A}\right)A^2 = A^3 + \frac{1}{\sqrt{2N_f}}\hat{A}A^2 \tag{2.7.4}$$

$$\mathcal{A}^5 = \left(A^3 + \frac{1}{\sqrt{2N_f}}\hat{A}A^2\right)A^2 = A^5 + \frac{1}{\sqrt{2N_f}}\hat{A}A^4 \tag{2.7.5}$$

$$\mathcal{F}^2 = F^2 + \frac{1}{\sqrt{2N_f}}(F\hat{F} + \hat{F}F) + \frac{1}{2N_f}\hat{F}^2 = F^2 + \frac{1}{\sqrt{2N_f}}\mathrm{d}\hat{A}F + \frac{1}{2N_f}\mathrm{d}\hat{A}^2 \tag{2.7.6}$$

$$\begin{aligned}\mathcal{A}^3\mathcal{F} &= \left(A^3 + \frac{1}{\sqrt{2N_f}}\hat{A}A^2\right)\left(F + \frac{1}{\sqrt{2N_f}}\hat{F}\right) \\ &= A^3F + \frac{1}{2N_f}\mathrm{d}\hat{A}A^3 + \frac{1}{\sqrt{2N_f}}\hat{A}A^2F + \frac{1}{2N_f}\hat{A}A^2\mathrm{d}\hat{A}\end{aligned} \tag{2.7.7}$$

$$\begin{aligned}\mathcal{A}\mathcal{F}^2 &= \left(A + \frac{1}{\sqrt{2N_f}}\hat{A}\right)\left(F^2 + \frac{1}{\sqrt{2N_f}}\mathrm{d}\hat{A}F + \frac{1}{2N_f}\mathrm{d}\hat{A}^2\right) \\ &= AF^2 + \frac{2}{2N_f}\mathrm{d}\hat{A}AF + \frac{1}{2N_f}\mathrm{d}\hat{A}^2A + \frac{1}{\sqrt{2N_f}}\hat{A}F^2 + \frac{1}{N_f}\hat{A}\mathrm{d}\hat{A}F + \frac{1}{2N_f\sqrt{2N_f}}\hat{A}\hat{F}^2\end{aligned} \tag{2.7.8}$$

for the CS term, we get

$$\begin{aligned}S_{CS} = \frac{N_c}{24\pi^2}\int_{\mathbb{R}^4\times[0,\infty)} \mathrm{Tr}_f\Bigg[&\left(AF^2 - \frac{i}{2}A^3F - \frac{!}{10}A^5\right) \\ &+ \frac{2}{\sqrt{2N_f}}\mathrm{d}\hat{A}^2\mathrm{tr}AF + \frac{1}{\sqrt{2N_f}}\hat{A}\mathrm{tr}F^2 + \frac{1}{2N_f\sqrt{2N_f}}\hat{A}\hat{F}^2 \times N_f \\ &- \frac{i}{2\sqrt{2N_f}}\mathrm{d}\hat{A}\mathrm{tr}A^3 - \frac{i}{2\sqrt{2N_f}}\mathrm{d}\hat{A}\mathrm{tr}A^2F\Bigg]\end{aligned} \tag{2.7.9}$$

Here, we calculate the total derivative term;

$$\frac{1}{\sqrt{2N_f}}\mathrm{d}\left(\hat{A}\mathrm{Tr}_f\left(2FA-\frac{i}{2}A^3\right)\right)=\frac{2}{2N_f}\mathrm{d}\hat{A}\mathrm{tr}AF-\frac{i}{2\sqrt{2N_f}}\mathrm{d}\hat{A}\mathrm{tr}A^3$$
$$-\frac{1}{\sqrt{2N_f}}\hat{A}\mathrm{tr}\left(2\mathrm{d}FA+2F\mathrm{d}A-\frac{i}{2}\mathrm{d}AA^2+\frac{i}{2}A\mathrm{d}AA-\frac{i}{2}A^2\mathrm{d}A\right) \tag{2.7.10}$$

Furthermore, the calculation of the term of the second line yields

$$\begin{aligned}
(\text{second line}) &= -\frac{2}{2N_f}\hat{A}\mathrm{tr}(-\mathrm{d}AA-iA\mathrm{d}A)A-\frac{2}{\sqrt{2N_f}}\hat{A}\mathrm{tr}F\mathrm{d}A+\frac{3i}{2\sqrt{2N_f}}\hat{A}\mathrm{tr}A^2\mathrm{d}A\\
&= -\frac{4i}{\sqrt{2N_f}}\hat{A}\mathrm{tr}A^2\mathrm{d}A-\frac{2}{\sqrt{2N_f}}\hat{A}\mathrm{tr}F\mathrm{d}A+\frac{3i}{2\sqrt{2N_f}}\hat{A}\mathrm{tr}A^2\mathrm{d}A\\
&= -\frac{5i}{2\sqrt{2N_f}}\hat{A}\mathrm{tr}A^2\mathrm{d}A-\frac{2}{\sqrt{2N_f}}\hat{A}\mathrm{tr}F\mathrm{d}A\\
&= -\frac{i}{2\sqrt{2N_f}}(\hat{A}\mathrm{tr}A^2\mathrm{d}A+\hat{A}\mathrm{tr}A^2iA^2)-\frac{1}{2\sqrt{2N_f}}\hat{A}\mathrm{tr}A^4-\frac{2i}{\sqrt{2N_f}}\hat{A}\mathrm{tr}A^2\mathrm{d}A\\
&\quad -\frac{2}{\sqrt{2N_f}}(\hat{A}\mathrm{tr}F\mathrm{d}A+\hat{A}\mathrm{tr}FiA^2)+\frac{2i}{\sqrt{2N_f}}\hat{A}\mathrm{tr}\mathrm{d}AA^2-\frac{2}{\sqrt{2N_f}}\hat{A}\mathrm{tr}A^4\\
&= -\frac{i}{2\sqrt{2N_f}}\hat{A}\mathrm{tr}A^2F-\frac{2}{\sqrt{2N_f}}\hat{A}\mathrm{tr}F^2,
\end{aligned} \tag{2.7.11}$$

which shows the desired expression.

2.7.2 Equation (2.4.20)

For $SU(2)$, by identity,

$$\omega_5(A)=\mathrm{Tr}\left(AF^2-\frac{i}{2}A^3F-\frac{1}{10}A^5\right)=0. \tag{2.7.12}$$

The formula for the product of generators of $SU(N)$ is listed below.

$$T^aT^b=\frac{1}{2N}\delta^{ab}+\frac{1}{2}d^{abc}T^c+\frac{i}{2}f^{abc}T^c \tag{2.7.13}$$

$$\mathrm{tr}(T^aT^b)=\frac{1}{2}\delta^{ab} \tag{2.7.14}$$

$$\mathrm{tr}(T^aT^bT^c)=\frac{1}{4}(d^{abc}+if^{abc}) \tag{2.7.15}$$

$$\mathrm{tr}(T^aT^bT^cT^d)=\frac{1}{4N}\delta^{ab}\delta^{cd}+\frac{1}{8}(d^{abe}+if^{abe})(d^{cde}+if^{cde}) \tag{2.7.16}$$

$$\mathrm{tr}(T^aT^bt^cT^dT^e)=\frac{1}{8N}\delta^{ab}(d^{cde}+if^{cde})+\frac{1}{4N}(d^{abe}+if^{abe})\delta^{cd}$$

$$+\frac{1}{16}(d^{abf}+id^{abf})(d^{cdg}+d^{cdg})(d^{efg}+d^{efg}) \tag{2.7.17}$$

Using the fact that, $d^{abc}=0$ for $N_f=2$, there are only three generators of SU(2), and that in the calculation of $\mathrm{tr}(A^5)$, the terms involving δ^{ab} are also zero due to the antisymmetry of the wedge product, we have $\omega_5(A)=0$ for SU(2).

2.7.3 Equation (2.4.22)

For arbitrary N_f, we show

$$S_{CS}=\frac{N_c}{24\pi^2}\sqrt{\frac{2}{N_f}}\int\left[\frac{3}{2}\hat{A}\mathrm{tr}F^2+\frac{1}{4}\hat{A}\hat{F}^2+\mathrm{d}(\ )\right]$$
$$\frac{N_c}{24\pi^2}\epsilon_{MNPQ}\int d^4xdz\Big[\frac{3}{8}\hat{A}_0\mathrm{Tr}_f(F_{MN}F_{PQ})-\frac{3}{2}\hat{A}_M\mathrm{Tr}_f(\partial_0A_NF_{PQ})$$
$$+\frac{3}{4}\hat{F}_{MN}\mathrm{Tr}_f(A_0F_{PQ})+\frac{1}{16}\hat{A}_0\hat{F}_{MN}\hat{F}_{PQ}-\frac{1}{4}\hat{A}_M\hat{F}_{0N}\hat{F}_{PQ}+(\text{total derivative term})\Big]. \tag{2.7.18}$$

If we factor out $\sqrt{2/N_f}$, this calculation can be used regardless of N_f. The point is to drop terms over two squares in the $SU(N_f)$ field (A and F). This is because considering $g_{YM}^2\ll 1/\lambda\ll 1$, g_{YM}^3 should be dropped, which is consistent with keeping the zeroth-order of λ in the analysis of baryons here.

Using that the differential form is written by

$$\omega_n=\frac{1}{n!}\mathrm{d}^nx\epsilon^{\mu_1\cdots\mu_n}\omega_{\mu_1\cdots\mu_n}, \tag{2.7.19}$$

we expand

$$\hat{A}\mathrm{tr}F^2=d^5x\epsilon^{0MNPQ}\frac{1}{2!}\frac{1}{2!}(\hat{A}_0\mathrm{tr}F_{MN}F_{PQ}-\hat{A}_M\mathrm{tr}F_{0N}F_{PQ}+\hat{A}_M\mathrm{tr}F_{N0}F_{PQ}$$
$$-\hat{M}\mathrm{tr}F_{NP}F_{0Q}+\hat{A}_M\mathrm{tr}F_{NP}F_{Q0})$$
$$=d^5x\epsilon^{0MNPQ}\frac{1}{4}(\hat{A}_0\mathrm{tr}F_{MN}F_{PQ}-\hat{A}_M\mathrm{tr}\partial_0A_NF_{PQ}\times 4+\hat{A}_M\mathrm{tr}\partial_NA_0F_{PQ}\times 4)+\mathcal{O}(A^3), \tag{2.7.20}$$

the last term becomes

$$\epsilon^{0MNPQ}\hat{A}_M\mathrm{tr}\partial_NA_0F_{PQ}=-\epsilon^{0MNPQ}(\partial_N\hat{A}_M\mathrm{tr}A_0F_{PQ}+\hat{A}_M\mathrm{tr}A_0\partial_NF_{PQ})$$
$$=-\epsilon^{0MNPQ}\partial_N\hat{A}_M\mathrm{tr}A_0F_{PQ}$$
$$=-\frac{1}{2}\epsilon^{0MNPQ}\hat{F}_{MN}\mathrm{tr}A_0F_{PQ}, \tag{2.7.21}$$

Therefore, we obtain

$$\hat{A}\mathrm{tr}F^2 = d^5x\epsilon^{0MNPQ}\Big(\frac{1}{4}\hat{A}_0\mathrm{tr}F_{MN}F_{PQ} - \hat{A}_M\mathrm{tr}\partial_0A_NF_{PQ} - \frac{1}{2}\hat{F}_{MN}\mathrm{tr}A_0F_{PQ}\Big) + \mathcal{O}(A^3). \tag{2.7.22}$$

Also, the term with only U(1) term expand

$$\hat{A}\hat{F}^2 = \epsilon^{0MNPQ}(\hat{A}_0\hat{F}_{MN}\hat{F}_{PQ} - 4\hat{A}_M\hat{F}_{0N}\hat{F}_{PQ}), \tag{2.7.23}$$

so the first equation could be proved.

2.7.4 Equation (2.4.49)

We prove

$$\begin{aligned} M =& 8\pi^2\kappa + \kappa\lambda^{-1}\int d^3xdz\Big[-\frac{z^2}{6}\mathrm{Tr}_f(F_{ij})^2 + z^2\mathrm{Tr}_f(F_{iz})^2\Big] \\ &-\frac{1}{2}\kappa\lambda^{-1}\int d^3xdz\Big[(\partial_M\hat{A}_0)^2 + \frac{1}{32\pi^2a}\hat{A}_0\epsilon_{MNPQ}\mathrm{Tr}_f(F_{MN}F_{PQ})\Big] + \mathcal{O}(\lambda^{-1}) \\ =& 8\pi^2\kappa\Big[1 + \lambda^{-1}\Big(\frac{\rho^2}{6} + \frac{1}{320\pi^4a^2}\frac{1}{\rho^2} + \frac{Z^2}{3}\Big) + \mathcal{O}(\lambda^{-1})\Big] \end{aligned} \tag{2.7.24}$$

For this, we can just integrate by substituting the classical solution obtained. One should use as follows;

$$F_{ij} = \frac{2\rho^2}{(\xi^2+\rho^2)^2}\epsilon_{ija}\tau^a, \quad F_{zj} = \frac{2\rho^2}{(\xi^2+\rho^2)^2}\tau_j. \tag{2.7.25}$$

2.7.5 Equation (2.4.60)

Calculate the field strength of the time-dependent gauge field in collective coordinates. By substituting the time-dependent gauge field into $F_{0M} = \partial_0A_M - \partial_MA_0 + i[A_0, A_M]$, the field strength is calculated as

$$\begin{aligned} F_{0M} =& \partial_0A_M - \partial_M\Delta A_0 + i[\Delta A_0, A_M] \\ =& \dot{W}A_M^{cl}W^{-1} + WA_M^{cl}\dot{W}^{-1} + W\partial_0A_M^0W^{-1} \\ &- W(\partial_MW^{-1}\Delta A_0W)W^{-1} - iW[A_M^{cl}, W^{-1}\Delta A_0W]W^{-1}. \end{aligned} \tag{2.7.26}$$

Here, using

$$\partial_0(W^{-1}W) = 0 = \dot{W}^{-1}W + W^{-1}\dot{W} \tag{2.7.27}$$

we calculate

$$\begin{aligned}
&\dot{W}A^{cl}_M W^{-1} + WA^{cl}_M \dot{W}^{-1} \\
&= W(W^{-1}\dot{W}A^{cl}_M)W^{-1} - W(A^{cl}_M W^{-1}\dot{W})W^{-1} \\
&= W[W^{-1}\dot{W}, A^{cl}_M]W^{-1} \\
&= Wi[A^{cl}_M, iW^{-1}\dot{W}]W^{-1}
\end{aligned} \tag{2.7.28}$$

and we obtain

$$\begin{aligned}
F_{0M} &= W\partial_0 A^0_M W^{-1} + W\partial_M(iW^{-1}\dot{W})W^{-1}W^{-1} + W(i[A^{cl}_M, iW^{-1}\dot{W}])W^{-1} \\
&\quad - WD^{cl}_M(W^{-1}\Delta A_0 W)W^{-1} \\
&= W\left(\dot{X}^\alpha \frac{\partial}{\partial X^\alpha}A^{cl}_M\right)W^{-1} - \left(WD^{cl}_M(W^{-1}\Delta A_0 W)W^{-1} - WD^{cl}_M(iW^{-1}\dot{W})W^{-1}\right) \\
&= W\left(\dot{X}^\alpha \frac{\partial}{\partial X^\alpha}A^{cl}_M - D^{cl}_M\Phi\right),
\end{aligned} \tag{2.7.29}$$

where Φ is defined by

$$\Phi = W^{-1}\Delta A_0 W - W^{-1}\dot{W} \tag{2.7.30}$$

With $\partial_M W(t) = 0$, note that we introduce, for example,

$$(0 =)W\partial_M(iW^{-1}\dot{W})W^{-1}W^{-1}. \tag{2.7.31}$$

2.7.6 Equation (2.4.67)

Here, we determine the metric of the collective coordinates,

$$L = \frac{m_X}{2}g_{\alpha\beta}\dot{X}^\alpha\dot{X}^\beta - U(X^\alpha) + \mathcal{O}(\lambda^{-1}) \tag{2.7.32}$$

$$U(X^\alpha) = U(\rho, Z) = M_0 + m_X\left(\frac{\rho^2}{6} + \frac{1}{320\pi^4 a^2}\frac{1}{\rho^2} + \frac{Z^2}{3}\right) \tag{2.7.33}$$

$$ds^2 = g_{\alpha\beta}dX^\alpha dX^\beta \tag{2.7.34}$$

$$= d\vec{X}^2 + dZ^2 + 2(d\rho^2 + \rho^2 da_I^2) \tag{2.7.35}$$

$$= d\vec{X}^2 + dZ^2 + 2dy^2. \tag{2.7.36}$$

To do this, we use

$$
\begin{aligned}
+ aN_c \int d^4xdz\mathrm{Tr}_f F_{0M}^2 &= aN_c \int d^4xdz\mathrm{Tr}_f (D_M^{cl}\Phi - \dot{A}_M^{cl})^2 \\
&= \frac{m_{X^\alpha}}{2} g_{\alpha\beta}\dot{X}^\alpha \dot{X}^\beta .
\end{aligned} \tag{2.7.37}
$$

Then, we need to solve

$$
D_M^{cl}\left(\dot{X}^\alpha \frac{\partial}{\partial X^\alpha} A_M^{cl} - D_M^{cl}\Phi\right) = 0. \tag{2.7.38}
$$

Since we desire a metric such that each of $(\vec{X}, Z, a_I)$ is diagonalized, we decompose as

$$
\Phi = \Phi_X + \Phi_\rho + \Phi_{SU(2)}. \tag{2.7.39}
$$

As a result, the equations we need to solve are also decoupled to become

$$
D_M^{cl}\left(\dot{X}^N \frac{\partial}{\partial X^N} A_M^{cl} - D_M^{cl}\Phi_X\right) = 0 \tag{2.7.40}
$$

$$
D_M^{cl}\left(\dot{\rho}\frac{\partial}{\partial\rho} A_M^{cl} - D_M^{cl}\Phi_\rho\right) = 0 \tag{2.7.41}
$$

$$
D_M^{cl} D_M^{cl}\Phi_{SU(2)} = 0. \tag{2.7.42}
$$

We will solve them in the steps below. It is useful to know that we can write

$$
g\partial_M g^{-1} = \begin{cases} \frac{i}{\xi^2}\left((z - Z)\tau^i - \epsilon_{ija}(x^j - X^j)\tau^a\right) \\ -\frac{i}{\xi^2}(x^a - X^a\tau^a \end{cases} \tag{2.7.43}
$$

$$
\partial_M(g\partial_M g^{-1}) \propto (x^M - X^M)g\partial_M g^{-1} = 0 \tag{2.7.44}
$$

for the BPST solution.

First, for

$$
D_M^{cl}\left(\dot{X}^N \frac{\partial}{\partial X^N} A_M^{cl} - D_M^{cl}\Phi_X\right) = 0, \tag{2.7.45}
$$

We easily find that we can write

$$
\Phi_X = -\dot{X}^N A_N^{cl} \tag{2.7.46}
$$

for the BPST solution. From this, noting

$$
\frac{\partial}{\partial_N} A_M^{cl} = -\partial_N A_M^{cl} \tag{2.7.47}
$$

we obtain

$$\dot{X}^N \frac{\partial}{\partial X^N} A_M^{cl} - D_M^{cl} \Phi_X = -\dot{X}^N \partial_N A_M^{cl} + \dot{X}^N \partial_M A_N^{cl} + i \dot{X}^N [A_M^{cl}, A_N^{cl}]$$
$$= \dot{X}^N F_{MN}^{cl}. \tag{2.7.48}$$

Therefore, we calculate

$$\frac{m_X}{2} g_{MN} \dot{X}^M \dot{X}^N = \kappa \lambda^{-1} \int d^3x dz \mathrm{tr}\big[(-\dot{X}^M F_{PM}^{cl})(-\dot{X}^N F_{PN}^{cl})\big]$$
$$= \kappa \lambda^{-1} \dot{X}^M \dot{X}^N \int d^3x dz \mathrm{tr} F_{PM}^{cl} F_{PN}^{cl}, \tag{2.7.49}$$

then, by substituting the solution and integrating this, we obtain

$$g_{MN} = \frac{2\kappa\lambda^{-1}}{m_X} \int d^3x dz \mathrm{tr} F_{MP}^{cl} F_{NP}^{cl} = \delta_{MN}. \tag{2.7.50}$$

Next, for

$$D_M^{cl} \Big(\dot{\rho} \frac{\partial}{\partial \rho} A_M^{cl} - D_M^{cl} \Phi_\rho \Big) = 0 \tag{2.7.51}$$

because of

$$\partial_\rho A_M^{cl} = -\frac{2\rho}{\xi^2 + \rho^2} A_M^{cl}, \tag{2.7.52}$$

we can calculate

$$D_M^{cl} \dot{\rho} \partial_\rho A_M^{cl} = \partial_M \Big(-\frac{2\dot{\rho}\rho}{\xi^2 + \rho^2} A_M^{cl} \Big) - \frac{2i\dot{\rho}\rho}{\xi^2 + \rho^2} [A_M^{cl}, A_M^{cl}] = 0, \tag{2.7.53}$$

then we obtain

$$\Phi_\rho = 0. \tag{2.7.54}$$

From this, in the same way as above, we get the following metric

$$g_{\rho\rho} = \frac{2\kappa\lambda^{-1}}{m_X} \int d^3x dz \mathrm{tr} \Big(\frac{\partial}{\partial \rho} A_M^{cl} \Big)^2 = 2. \tag{2.7.55}$$

Finally, for

$$D_M^{cl} D_M^{cl} \Phi_{SU(2)} = 0, \tag{2.7.56}$$

Moving on to singular gauge,

$$\overline{A}_M^{cl}(x) = g(x)^{-1}\big(A^{cl}(x) - i\partial_M\big)g(x) = -i(1 - f(\xi))g(x)^{-1}\partial_M g(x) \quad (2.7.57)$$

$$\overline{\Phi}_{SU(2)}(t, x) = g(x; X(t))^{-1}\Phi_{SU(2)}(t, x)g(x; X(t)) \quad (2.7.58)$$

we solve

$$\overline{D}_M^{cl}\overline{D}_M^{cl}\overline{\Phi}_{SU(3)} = 0. \quad (2.7.59)$$

Assuming the ansatz;

$$\overline{\Phi}_{SU(3)} = \sum_{a=1}^{3} \chi^a(t)u^a(\xi)\frac{\tau_a}{2} \quad (2.7.60)$$

the equation we have to now solve becomes

$$\frac{1}{\xi^3}\frac{d}{d\xi}\left(\xi^3\frac{d}{d\xi}u^a(\xi)\right) = 8\frac{(1 - f(\xi))^2}{\xi^2}u^a(\xi) \quad (2.7.61)$$

with arbitrary constants C^a, the regular solution is obtained as

$$u^a(\xi) = C^a f(\xi). \quad (2.7.62)$$

Moving this back to the previous gauge, then we obtain

$$\Phi_{SU(2)} = \chi^a(t)\Phi_a(x) \quad (2.7.63)$$

$$\Phi_a = f(\xi)g\frac{\tau_a}{2}g^{-1} \quad (2.7.64)$$

The arbitrary constant C^a was absorbed in χ^a. Still $\chi^a(t)$ has not been determined, but if we remind that

$$\overline{\mathcal{A}}(t, x^M) = \left(\overline{\mathcal{A}}^{cl}(x^M; X^a(t)) + \overline{\Phi}(t, x^M)dt\right)^{W(t)} \quad (2.7.65)$$

$$\overline{\Phi}(t, x^M) = g(x^M; X^\alpha(t))\big(\Phi(t, x^M) - i\partial_0\big)g(x^M; X^\alpha(t))^{-1} \quad (2.7.66)$$

is gauge-fixed such that it is regular at infinity, even after giving it a time dependence, we should still require

$$\overline{\mathcal{A}}(t, x^M) \to 0. \quad (2.7.67)$$

Therefore, with $\Phi = \Phi_X + \Phi_\rho + \Phi_{SU(2)}$, at infinity, $\overline{\mathcal{A}}^{cl} \to 0$, so we should require

$$W\overline{\Phi}W^{-1} - iW\partial_0 W^{-1} \to 0. \quad (2.7.68)$$

Then, χ^a are determined as

$$\chi^a(t) = -2\text{tr}(t_a W^{-1}\dot{W}) = 2(a_4\dot{a}_a - \dot{a}_4 a_a + \epsilon_{abc}a_b\dot{a}_c), \quad t_a = \frac{\tau_a}{2}. \tag{2.7.69}$$

From the above, we obtain $\Phi_{S(2)}$. From this, the metric are given as

$$g_{IJ}\dot{a}_I\dot{a}_J = \frac{2\kappa\lambda^{-1}}{m_X}\int d^3x dz \text{tr}\left(D_M^{cl}\Phi_{SU(2)}\right)^2 = 2\rho^2\dot{a}_I^2, \tag{2.7.70}$$

where we use

$$(\chi^a)^2 = 4\dot{a}_I^2 \tag{2.7.71}$$

$$a_I^2 = 1. \tag{2.7.72}$$

After the above calculations, we get

$$S = \int dt L = \int dt(L_{\text{kinetic}} - L_{\text{potential}}) \tag{2.7.73}$$

$$L = L_X + L_Z + L_y + \mathcal{O}(\lambda^{-1}) \tag{2.7.74}$$

$$L_X = -M_0 + \frac{m_X}{2}\dot{\mathbf{X}}^2 \tag{2.7.75}$$

$$L_Z = \frac{m_Z}{2}\dot{Z}^2 - \frac{m_Z\omega_Z^2}{2}Z^2 \tag{2.7.76}$$

$$L_y = \frac{m_y}{2}(\dot{\rho}^2 + \rho^2\dot{a}_I^2) - \frac{m_y\omega_\rho^2}{2}\rho^2 - \frac{Q}{\rho^2} \tag{2.7.77}$$

$$= \frac{m_y}{2}\dot{y}_I^2 - \frac{m_y\omega_\rho^2}{2}y_I^2 - \frac{Q}{\rho^2} \tag{2.7.78}$$

with coefficients

$$M_0 = 8\pi^2\kappa, \quad m_X = m_Z = \frac{m_y}{2} = 8\pi^2\kappa\lambda^{-1} = 8\pi^2 a N_c$$
$$\omega_Z^2 = \frac{2}{3}, \quad \omega_\rho^2 = \frac{1}{6}, \quad Q = \frac{N_c^2}{5m_X} = \frac{N_c}{40\pi^2 a}. \tag{2.7.79}$$

References

1. Tadakatsu S, Shigeki S (2005) Low energy hadron physics in holographic QCD. Prog Theor Phys 113:843–882
2. Tadakatsu S, Shigeki S (2005) More on a holographic dual of QCD. Prog Theor Phys 114:1083–1118
3. Joseph P (1995) Dirichlet branes and Ramond-Ramond charges. Phys Rev Lett 75:4724–4727
4. Juan Martin Maldacena (1998) The Large N limit of superconformal field theories and supergravity. Adv Theor Math Phys 2:231–252
5. Andreas K, Emanuel K (2002) Adding flavor to AdS/CFT. JHEP 06:043

6. Edward W (1998) Anti-de sitter space, thermal phase transition, and confinement in gauge theories. Adv Theor Math Phys 2:505–532
7. Son DT, Stephanov MA (2004) QCD and dimensional deconstruction. Phys Rev D 69:065020
8. 村田 仁樹., Hadrons in holographic QCD, 京都大学大学院理学研究科　修士論文.
9. Yizhuang L, Ismail Z (2017) Holographic heavy-light chiral effective action. Phys Rev D 95(5):056022
10. Edward W (1998) Baryons and branes in anti-de Sitter space. JHEP 07:006
11. Douglas Michael R (1999) Branes within branes. NATO Sci Ser C 520:267–275
12. Atiyah MF, Manton NS (1989) Skyrmions from instantons. Phys Lett B 222:438–442
13. Kanabu N, Hideo S, Toru K (2007) Baryons in holographic QCD. Phys Rev D 75:086003
14. Hiroyuki H, Tadakatsu S, Shigeki S, Shinichiro Y (2007) Baryons from instantons in holographic QCD. Prog Theor Phys 117:1157
15. Gervais JL, Sakita B (1975) Extended particles in quantum field theories. Phys Rev D 11:2943
16. Manton NS (1982) A remark on the scattering of BPS monopoles. Phys Lett B 110:54–56
17. Myers RC (1999) Dielectric branes. JHEP 12:022
18. Adkins Gregory S, Nappi Chiara R, Edward W (1983) Static properties of nucleons in the skyrme model. Nucl Phys B 228:552
19. Adkins Gregory S, Nappi Chiara R (1984) Stabilization of chiral solitons via vector mesons. Phys Lett B 137:251–256
20. Belavin AA, Polyakov AM, Schwartz AS, Tyupkin YS (1975) Pseudoparticle solutions of the Yang-Mills equations. Phys Lett B 59:85–87
21. Guadagnini E (1984) Baryons as solitons and mass formulae. Nucl Phys B 236:35–47
22. Yabu H, Ando K (1988) A new approach to the SU(3) skyrme model. Nucl Phys B 301:601
23. Sanjay J, Wadia Spenta R (1985) Large-N baryons: collective coordinates of the topological soliton in the SU(3) chiral model. Nucl Phys B 258:713–725
24. Edward W (1979) Baryons in the 1/n expansion. Nucl Phys B 160:57–115
25. Ammon M, Erdmenger J (2015) Gauge/gravity duality: foundations and applications, vol 4. Cambridge University Press, Cambridge
26. Manton NS, Sutcliffe P (2004) Topological solitons. Cambridge monographs on mathematical physics. Cambridge University Press, Cambridge

Chapter 3
SU($N_f = 2 + 1$) Sakai-Sugimoto Model

Abstract We introduce a strange quark in the Sakai-Sugimoto model by collective rotation. For easier understanding, we first explain this concept in terms of Skymion, where this method has traditionally been used. We review the problem of the Chern-Simons term that arises when this method is applied to the Sakai-Sugimoto model, and how to solve it. Finally, using the dimensional reduction method proposed by Forgacs-Manton, we show one way to introduce charm and bottom quarks into the model by utilizing the extra dimensions of this model.

In this chapter, we discuss the description of hadrons by the $SU(N_f = 2 + 1)$ Sakai-Sugimoto model toward the analysis of hyperons and heavy baryons. The s quarks can be introduced by SU(3) rotations in the collective coordinate quantization, as is accomplished in the conventional quantization of solitons. To demonstrate this, Sect. 3.1 will review how to introduce s-quarks as SU(3) rotations in the Skyrme model. Then, in Sect. 3.2, we review how to introduce s-quarks using the same method in the Sakai-Sugimoto model. At this time, it is known that the following two problems occur.

1. Although the introduction of the mass of the s-quark is important for the analysis of hyperons, the Sakai-Sugimoto model is a massless QCD, which does not have an established way to introduce the mass.
2. In flavor SU(3), the constraint term that should be imposed on the hypercharge, which is necessary to consider solitons as baryons (fermions), does not appear from the Chern Simons term (CS term) of the Sakai-Sugimoto model.

For the first point, we will briefly review one way to introduce mass. For the second point, we succeeded in obtaining a constraint term in Ref. [1], but the chiral anomaly, which was well reproduced in the original Sakai-Sugimoto model, disappeared. Later, in Ref. [2], the CS term was obtained that reproduced the chiral anomaly and still produced the constraint term. These points are discussed in detail in Appendix 3.6 because they are beyond the scope of this thesis.

D. Fujii, *Dynamical Properties of Baryon Resonances in the Holographic QCD*,
Springer Theses, https://doi.org/10.1007/978-981-97-0755-3_3

Next, we show how to introduce heavy quarks such as c and b into the Sakai-Sugimoto model, based on the method we proposed in Ref. [3]. We obtain the mass spectrum of the heavy baryons by collective coordinate quantization of the classical solution of the action obtained using the dimensional reduction method of gauge theory proposed by Forgács-Manton. This method, which is similar to the bound state approach in the Skyrme model [4–8], is shown to reproduce well the mass spectra of the heavy baryon and its resonance states, as well as the P_c state, which has attracted much attention due to its recent experimental observations. Similar studies have been made in Ref. [9–14].

In this thesis, we will explain the following flow concerning the study of the Sakai-Sugimoto model of $SU(N_f = 2 + 1)$. In Sect. 3.1, we first explain the flavor SU(3) Skyrme model to prepare the analysis of baryons in the flavor SU(3) Sakai-Sugimoto model. The necessity of the constraint term and the analysis method when a mass term is incorporated will be useful for understanding subsequent sections. In Sect. 3.2, we discuss how to introduce s-quarks in the Sakai-Sugimoto model. The problems with the constraint terms are described in Appendix 3.6. It can be seen that this constraint term does not appear from the CS term used in the original paper on the Sakai-Sugimoto model [15, 16]. We explained in the Appendix along with Ref. [1, 2], which solved this problem. In the former paper, the chiral anomaly disappeared, and the latter paper realized both the constraint term and the chiral anomaly. Then, in the Sect. 3.2.2, we will briefly review how to deform the Sugimoto model so that quarks have masses [17–19]. Next, we discuss how to introduce heavy quarks such as c and b quarks into the Sakai-Sugimoto model. For this purpose, the method of dimensional reduction proposed by Forgács-Manton will be explained. Then, we introduce heavy quarks into the Sakai-Sugimoto model and obtain their mass spectra. This discussion is based on our study Ref. [3].

3.1 Flavor SU(3) Skyrme Model

In this section we explain how to obtain the Hamiltonian of a quantum mechanical system in collective coordinates through the flavor SU(3) Skyrme model. This explanation helps to understand similar arguments in the Sakai-Sugimoto model (the difference with the Sakai-Sugimoto model is the absence or not of (ρ, Z)). It is important in the next section to discuss how the constraint terms impose restrictions on the state and how to deal with the mass terms. In this section, we have adapted the notation to Ref. [20] and discuss it in more detail with reference to Ref. [21] etc.

3.1.1 Classical Solutions

The action of the SU(3) Skyrme model is written by

$$S = S_{\mathrm{Sk}} + S_{\mathrm{SB}} + S_{\mathrm{WZW}} \tag{3.1.1}$$

$$S_{\mathrm{Sk}} = \int d^4x\Big(\frac{f_\pi^2}{4}\mathrm{Tr}\partial_\mu U\partial^\mu U^\dagger + \frac{1}{32e^2}\mathrm{Tr}[U^\dagger\partial_\mu U, U^\dagger\partial_\nu U]^2\Big) \tag{3.1.2}$$

$$\begin{aligned} S_{\mathrm{SB}} = &\int d^4x\Big(\frac{f_\pi^2}{4}(m_\pi^2 + m_\eta^2)\mathrm{Tr}(U + U^\dagger - 2) \\ &+ \frac{\sqrt{3}}{6}f_\pi^2(m_\pi^2 - m_K^2)\mathrm{Tr}[\frac{1}{2}t_8(U + U^\dagger)]\Big) \end{aligned} \tag{3.1.3}$$

$$\begin{aligned} S_{\mathrm{WZW}} &= -\frac{N_c}{240\pi^2}\int_{S^5}\mathrm{Tr}[U^\dagger dU]^5 \\ &= -\frac{N_c}{240\pi^2}\epsilon^{\alpha\beta\gamma\delta\epsilon}\int_{S^3\times D_2} d^5x\mathrm{Tr}[U^\dagger\partial_\alpha UU^\dagger\partial_\beta UU^\dagger\partial_\gamma UU^\dagger\partial_\delta UU^\dagger\partial_\epsilon U], \end{aligned} \tag{3.1.4}$$

where, with $x^\alpha = t, x^1, x^2, x^3, s$, the 2-dimensional disk D_2 is parameterized by (t, s). The first term in (3.1.2) is a chiral term, $\mu = 0, 1, 2, 3$ $U(\mathbf{x}) = e^{4i\pi_a(\mathbf{x})t_a/f_\pi}$, $\mathrm{tr}(t_at_b) = \delta_{ab}/2$, and the second term is the Skyrme term, which is the higher-order derivative term needed to prevent the soliton from shrinking. The action (3.1.3) is the symmetry breaking term, which is expanded to lead to

$$L_{\mathrm{SB}} \simeq \int d^4x\Big(\frac{1}{2}m_\pi^2(\pi_1^2 + \pi_2^2 + \pi_3^2) + \frac{1}{2}m_K^2(\pi_4^2 + \pi_5^2 + \pi_6^2 + \pi_7^2) + \frac{1}{2}m_\eta^2\pi_8^2\Big). \tag{3.1.5}$$

The action (3.1.4) is the Wess-Zumino-Witten term, which is necessary to reproduce the chiral symmetry and is also important in the sense that only the parity symmetry in QCD is extracted from the redundant symmetry [22, 23].

As a static (time-independent) soliton solution $U_0(x)$ of baryon number 1, it is known that the SU(3) hedgehog solution given by

$$U_0(\mathbf{x}) = \begin{pmatrix} \exp(i\hat{x}\cdot F(r)) & 0 \\ 0 & 1 \end{pmatrix} \tag{3.1.6}$$

$$\hat{x} = \frac{\mathbf{x}}{r}, \quad F(r=0) = 0, \quad F(r\to\infty) = n\pi(=\pi). \tag{3.1.7}$$

3.1.2 Collective Coordinates

Let us give time dependence to the collective coordinates as in the previous section. Focusing only on the collective coordinates that generate the soliton rotation, we consider the motion in the moduli space as written as

$$U(t, \mathbf{x}) = W(t)U_0(t, \mathbf{x})W(t)^{-1} \tag{3.1.8}$$

$$W(t) = e^{i\xi_a(t)t_a}, \quad a = 1 \sim 8, \tag{3.1.9}$$

where, for S_{WZW} on the different manifold, it is defined as

$$W(t, s = 0) = W(t) \tag{3.1.10}$$

$$W(t, s = 1) = 1. \tag{3.1.11}$$

In fact, $W(t) \in SU(3)$ is not precisely a collective coordinate due to the symmetry breaking term, but we will treat it as a collective coordinate in the same sense as (ρ, Z) in the previous chapter.

The transformation of the collective coordinate $W(t)$ and the corresponding quantum numbers are given as follows;

1. Isospin rotation $SU(2)_L \subset SU(3)_L \ : \ (I, I_3)$

$$W(t) \to g_L W(t), \quad g_L \in SU(2)_L \tag{3.1.12}$$

2. Hypercharge $U(1)_L \subset SU(3)_L \ : \ Y$

$$W(t) \to e^{i\xi_L t_8} W(t), \quad e^{i\xi_L t_8} W(t) \in U(1)_L \tag{3.1.13}$$

3. spin rotation (spatial rotation) $SU(2)_R \subset SU(3)_R \ : \ (J, J_3)$

$$W(t) \to W(t) g_R^{-1}, \quad g_R \in SU(2)_R \tag{3.1.14}$$

4. Right hypercharge $U(1)_R \subset SU(3)_R \ : \ Y_R$

$$W(t) \to W(t) e^{-i\xi_R t_8} \in U(1)_R \tag{3.1.15}$$

$$Y_R = \frac{1}{3}\begin{pmatrix} 1 & 0 & 0 \\ 0 & 1 & 0 \\ 0 & 0 & -2 \end{pmatrix}, \quad Y_R = \frac{2}{\sqrt{3}} t_8 \tag{3.1.16}$$

It can be seen from

$$
\begin{aligned}
U(t,x) \to & W(t)e^{-i\xi_R t_8}\begin{pmatrix}\exp(i\hat{x}\cdot F(r)) & 0\\ 0 & 1\end{pmatrix}e^{i\xi_R t_8}W(t)^{-1}\\
&= U(t,x)
\end{aligned}
\tag{3.1.17}
$$

that the transformation of the right hypercharge is an unnecessary degree of freedom. Thus the moduli space is the homogeneous space $SU(3)/U(1)_R$, and Y_R is not a physical degree of freedom. We consider SU(3) as a moduli space according to Ref. [21, 24–27] instead of $SU(3)/U(1)_R$, and treat the degrees of freedom of $U(1)_R$ as constraints.

By substituting (3.1.8) into the action S (L_{SB} will be treated as perturbative later), the Lagrangian is given as

$$
S = \int dt L \tag{3.1.18}
$$

$$
\begin{aligned}
L =& - M_0 + \frac{1}{2}\mathcal{I}_0(\dot{a}_1^2 + \dot{a}_2^2 + \dot{a}_3^2) + \frac{1}{2}\mathcal{I}_0'(\dot{a}_4^2 + \dot{a}_5^2 + \dot{a}_6^2 + \dot{a}_7^2)\\
&+ \frac{2}{\sqrt{3}}N_c B(U_0)\dot{a}_8,
\end{aligned}
\tag{3.1.19}
$$

where $B(U_0)$ is the baryon number, $\mathcal{I}_0$ and $\mathcal{I}_0'$ are the coefficients obtained by performing the spatial integration numerically. Since we are now considering the case of baryon number 1, the Lagrangian can also be written as

$$
\begin{aligned}
L =& - M_0 + \frac{1}{2}\mathcal{I}_0\sum_{b=1}^{3}\dot{\xi}_a C_{ab}\dot{\xi}_c C_{cb} + \frac{1}{2}\mathcal{I}_0'\sum_{b=4}^{7}\dot{\xi}_a C_{ab}\dot{\xi}_c C_{cb}\\
&+ \frac{1}{2\sqrt{3}}N_c\dot{\xi}_a C_{a8},
\end{aligned}
\tag{3.1.20}
$$

with

$$
W(t)^{-1}\dot{W}(t) = iC_{ab}(\xi(t))t_a\dot{\xi}_b = it_a\dot{a}_a \tag{3.1.21}
$$

$$
\dot{a}_a^2 = 4[\mathrm{Tr}(-it_a W(t)^{-1}\dot{W}(t))]^2, \quad C_{ab}\dot{\xi}_b = \dot{a}_a \tag{3.1.22}
$$

$$
\left(\frac{4}{\sqrt{3}}N_c B(U_0)\mathrm{Tr}(-it_8 W^{-1}\dot{W}(t)) = \frac{2}{\sqrt{3}}N_c B(U_0)\dot{a}_8\right) \tag{3.1.23}
$$

3.1.3 Hamiltonian and Constraint Term

Let us quantize and find the Hamiltonian. From $L(\xi, \dot{\xi})$ (3.1.20), the canonical momentum of ξ is derived as

$$\pi_a = \frac{\partial L(\xi, \dot{\xi})}{\partial \dot{\xi}_a} = \mathcal{I}_0 \sum_{b=1}^{3} C_{ab}\dot{\xi}_c C_{cb} + \mathcal{I}_0' \sum_{b=4}^{7} C_{ab}\dot{\xi}_c C_{cb} + \frac{N_c}{2\sqrt{3}} C_{a8} \tag{3.1.24}$$

$$C_{ba}^{-1}\pi_a = \mathcal{I}_0 \dot{\xi}_a C_{ab}, \quad b = 1, 2, 3 \tag{3.1.25}$$

$$C_{ba}^{-1}\pi_a = \mathcal{I}_0' \dot{\xi}_a C_{ab}, \quad b = 4, 5, 6, 7 \tag{3.1.26}$$

$$C_{ba}^{-1}\pi_a = \frac{N_c}{2\sqrt{3}}, \quad b = 8, \tag{3.1.27}$$

where we impose the canonical commutation relation

$$[\xi_a, \pi_b] = i\delta_{ab} \tag{3.1.28}$$

$$[\xi_a, \xi_b] = 0, \quad [\pi_a, \pi_b] = 0.$$

We now define

$$R_a = C_{ba}^{-1}\pi_a. \tag{3.1.29}$$

This is the generator of $SU(3)_R$ acting on $W(t)$ from the right, as shown soon. In the same way as $C_{ab}(\xi)$, from

$$W(t)\dot{W}(t)^{-1} = iE_{ab}(\xi)t_a\dot{\xi}_b, \tag{3.1.30}$$

the $SU(3)_L$ generator is defined to be

$$L_a = E_{ba}^{-1}\pi_a. \tag{3.1.31}$$

From these definitions, we obtain the relation between R_a and L_a as

$$WW^{-1}\dot{W}W^{-1} = iC_{ab}(\xi)Wt_aW^{-1}\dot{\xi}_b$$

$$= -W\dot{W}^{-1} = -iE_{ab}(\xi)t_a\dot{\xi}_b$$

$$C_{ab}(\xi)Wt_aW^{-1} = -E_{ab}(\xi)t_a$$

$$L_a = -WR_aW^{-1}. \tag{3.1.32}$$

From the above, the two Casimir operators are

$$\mathrm{tr}[(R_a t_a)^2] = \mathrm{tr}[(L_a t_a)^2] \tag{3.1.33}$$

$$\mathrm{tr}[(R_a t_a)^3] = -\mathrm{tr}[(L_a t_a)^3], \tag{3.1.34}$$

which are interpreted as spin and isospin rotations, where they are complex conjugate representations of each other.

From the canonical commutation relation (3.1.28), we can derive

$$\pi_a = \frac{1}{i}\frac{\partial}{\partial \xi_a} \tag{3.1.35}$$

$$R_a = \frac{1}{i}C^{-1}_{ab}\frac{\partial}{\partial \xi_b}, \quad L_a = \frac{1}{i}E^{-1}_{ab}\frac{\partial}{\partial \xi_b}, \tag{3.1.36}$$

so by using (3.1.32), we obtain

$$\begin{aligned} W^{-1}(R_a W) =& \frac{1}{i}C^{-1}_{ab}W^{-1}\frac{\partial W}{\partial \xi_b} = \frac{1}{i}C^{-1}_{ab}iC_{bc}t_c \\ =& t_a \\ R_a W =& W t_a. \end{aligned} \tag{3.1.37}$$

Similarly, using (3.1.32), we obtain

$$\begin{aligned} WW^{-1}(L_a W)W^{-1} =& \frac{1}{i}WE^{-1}_{ab}W^{-1}\frac{\partial W}{\partial \xi_b}W^{-1} = \frac{1}{i}E^{-1}_{ab}iC_{bc}Wt_cW^{-1} \\ =& t_a \\ L_a W =& -t_a W. \end{aligned} \tag{3.1.38}$$

Furthermore, since

$$\begin{aligned} W^{-1}[R_a.R_b]W =& W^{-1}R_aWt_b - W^{-1}R_bWt_a = W^{-1}W[t_a,t_b] = if_{abc}t_c \\ =& if_{abc}W^{-1}R_aW \end{aligned} \tag{3.1.39}$$

$$\begin{aligned} W^{-1}[L_a.L_b]W =& -W^{-1}L_at_bW + W^{-1}L_bt_aW = -W^{-1}[t_a,t_b]W = -if_{abc}W^{-1}t_cW \\ =& if_{abc}W^{-1}L_aW \end{aligned} \tag{3.1.40}$$

and

$$\begin{aligned} R_aL_bW &= -t_bR_aW = -t_bWt_a \\ L_bR_aW &= L_aWt_a = -t_bWt_a \end{aligned} \tag{3.1.41}$$

are confirmed, we obtain the commutation relation between R_a and L_a,

$$[R_a.R_b] = if_{abc}R_a \tag{3.1.42}$$

$$[L_a.L_b] = if_{abc}L_a \tag{3.1.43}$$

$$[R_a, L_b] = 0. \tag{3.1.44}$$

From this commutation relation and (3.1.37) and (3.1.38), we identify R_a as the spin rotation of $SU(3)_R$ and L_a as the isospin rotation of $SU(3)_L$.

Using R_a (3.1.29), Hamiltonian can be expressed as

$$H = M_0 + \frac{1}{2}\frac{1}{\mathcal{I}_0}\sum_{b=1}^{3} R_a^2 + \frac{1}{2}\frac{1}{\mathcal{I}'_0}\sum_{b=4}^{7} R_a^2 \tag{3.1.45}$$

or

$$H = M_0 + \frac{1}{2}\left(\frac{1}{\mathcal{I}_0} - \frac{1}{\mathcal{I}'_0}\right)\sum_{b=1}^{3} R_a^2 + \frac{1}{2}\frac{1}{\mathcal{I}'_0}\sum_{b=1}^{8} L_a^2 - \frac{1}{2}\frac{1}{\mathcal{I}'_0}R_8^2. \tag{3.1.46}$$

Also, from (3.1.29) and (3.1.24) we have

$$\begin{aligned} Y_R &= \frac{2}{\sqrt{3}}R_8 = \frac{2}{\sqrt{3}}C_{8a}^{-1}\pi_a = \frac{2}{\sqrt{3}}\frac{N_c}{2\sqrt{3}} \\ &= \frac{N_c}{3}. \end{aligned} \tag{3.1.47}$$

In $N_c = 3$, the constraint $Y_R = 1$ is imposed on this system.

Since the weight diagram for the **8**- and **10**-dimensional representations of $SU(3)_R$ is Fig. 3.1, this constraint restricts the spin state to a half-integer (1/2 in the **8**-dimensional representation and 3/2 in the **10**-dimensional representation). These represent the spin 1/2 flavor octuplet baryon (N, Σ, Λ, Ξ) and the spin 3/2 flavor decuplet baryon (Δ, Σ^*, Ξ^*, Ω), respectively.

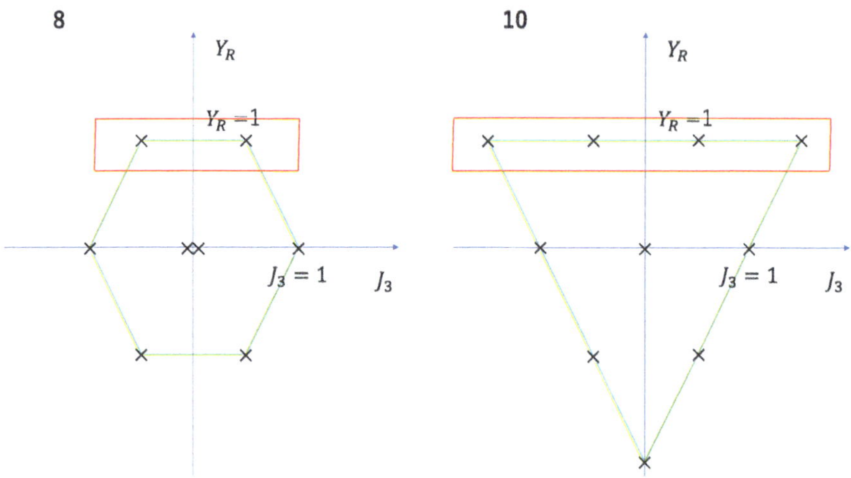

Fig. 3.1 Weight diagram and constraint

Here, considering (3.1.37) and (3.1.38), we define the wave function of the Hamiltonian (3.1.46) as a function as follows;

$$\Psi^{(p,q)}_{I,I_3,Y;J,J_3,Y_R=1}(W) = \langle (p,q); I, I_3, Y | D^{(n)}(W) | (p,q); J, J_3, Y_R = 1 \rangle, \quad (3.1.48)$$

where $D^{(n)}(W \in SU(3))$ is the $\boldsymbol{n}$-dimensional matrix representation of SU(3), and $SU(3)_L$ and $SU(3)_R$ generators, which were previously denoted t_a, are now distinguished as I_a and J_a (t_8 remains the same Y and Y_R). This function is called the Wigner D-function. Then, from (3.1.37) and (3.1.38), R_a and L_a act on the wave function as

$$R_a \Psi^{(p,q)}_{I,I_3,Y;J,J_3,Y_R=1}(W) = \langle (p,q); I, I_3, Y | D^{(n)}(W) J_a | (p,q); J, J_3, Y_R = 1 \rangle \quad (3.1.49)$$

$$L_a \Psi^{(p,q)}_{I,I_3,Y;J,J_3,Y_R=1}(W) = \langle (p,q); I, I_3, Y | I_a D^{(n)}(W) | (p,q); J, J_3, Y_R = 1 \rangle, \quad (3.1.50)$$

where the operations of J_a and I_a on $|(p,q); J, J_3, Y_R = 1\rangle$ and $\langle (p,q); I, I_3, Y|$ can be expressed by R_a, L_a and $\Psi^{(p,q)}_{I,I_3,Y;J,J_3,Y_R=1}(W)$. In other words, I_a and J_a act in Hilbert space, while L_a and R_a act in a space spanned by the coordinate ξ, where

$$\sum_{a=1}^{8} L_a^2 = \sum_{a=1}^{8} R_a^2 = \frac{1}{3}[p^2 + q^2 + pq + 3(p+q)] \quad (3.1.51)$$

$$\sum_{a=3}^{3} R_a^2 = J(J+1) \quad (3.1.52)$$

holds.

Finally, from this constraint, we show that the transformation properties of the wave function for spatial rotation are that of a fermion. For a spatial rotation, $W(t)$ transforms to

$$W(t) \to W(t) \begin{pmatrix} e^{i\theta/2} & 0 & 0 \\ 0 & e^{-i\theta/2} & 0 \\ 0 & 0 & 1 \end{pmatrix}, \quad \begin{pmatrix} e^{i\theta/2} & 0 \\ 0 & e^{-i\theta/2} \end{pmatrix} \in SU(2). \quad (3.1.53)$$

In particular, for the transformation of $\theta = 2\pi$ that makes one rotation, it transforms to

$$\begin{aligned} W(t) \to & W \begin{pmatrix} -1 & 0 & 0 \\ 0 & -1 & 0 \\ 0 & 0 & 1 \end{pmatrix} = W \exp\left[i\pi \begin{pmatrix} 1 & 0 & 0 \\ 0 & 1 & 0 \\ 0 & 0 & -2 \end{pmatrix} \right] \\ & = W e^{3i\pi Y_R} \end{aligned} \quad (3.1.54)$$

If we let it act on the wave function, it becomes

$$\begin{aligned} e^{3i\pi Y_R}\Psi^{(p,q)}_{I,I_3,Y;J,J_3,Y_R=1}(W) &= e^{3i\pi N_c/3}\langle (p,q); I, I_3, Y|D^{(n)}(W)|(p,q); J, J_3, Y_R = 1\rangle \\ &= e^{i\pi N_c}\Psi^{(p,q)}_{I,I_3,Y;J,J_3,Y_R=1}(W) \end{aligned} \tag{3.1.55}$$

so we know that it converts to

$$\Psi(W) \to e^{i\pi N_c}\Psi(W) = \begin{cases} -\Psi(W), & N_c = \text{odd} \\ +\Psi(W), & N_c = \text{even} \end{cases} \tag{3.1.56}$$

Therefore, we conclude that for $N_c = 3$, the wave function is a fermion.

3.1.4 Mass Deformation

In the previous discussions, the mass term

$$\begin{aligned} S_{SB} &= \int d^4x \left(\frac{f_\pi^2}{4}(m_\pi^2 + m_\eta^2)\mathrm{Tr}(U + U^\dagger - 2) \right. \\ &\quad \left. + \frac{\sqrt{3}}{6} f_\pi^2 (m_\pi^2 - m_K^2)\mathrm{Tr}\left[\frac{1}{2}t_8(U + U^\dagger)\right]\right) \\ &= -\frac{1}{2}\gamma\big(1 - D^{(8)}_{88}(W)\big) \end{aligned} \tag{3.1.57}$$

$$D^{(8)}_{88}(W) = \frac{1}{8}\mathrm{tr}(t_8 W^\dagger t_8 W) \tag{3.1.58}$$

has been neglected. In this thesis, we treat the mass term as a perturbation according to Ref. [21, 24–27] etc. Now the Hamiltonian is the sum of

$$\begin{aligned} H &= M_0 + \frac{1}{2}\left(\frac{1}{\mathcal{I}_0} - \frac{1}{\mathcal{I}'_0}\right)\sum_{b=1}^{3} R_a^2 + \frac{1}{2}\frac{1}{\mathcal{I}'_0}\sum_{b=1}^{8} L_a^2 - \frac{1}{2}\frac{1}{\mathcal{I}'_0}R_8^2 \\ \delta H &= +\frac{1}{2}\gamma\big(1 - D^{(8)}_{88}(W)\big). \end{aligned} \tag{3.1.59}$$

The problem we are considering now is the case $\delta H = \frac{1}{2}\gamma\big(1 - D^{(8)}_{88}(W)\big)$, while we generalize to $\delta H = \frac{1}{2}\gamma\big(1 - D^{(8)}_{ab}(W)\big)$, $(a = 1 \sim 8)$. With baryons labeled

$$B = (P, N, \Lambda, \Sigma^+, \Sigma^0, \Sigma^-, \Xi^0, \Xi^-, \Delta^{++}, \Delta^+, \Delta^0, \Delta^-, \Sigma^{*+}, \Sigma^{*0}, \Sigma^{*-}, \Xi^{*0}, \Xi^{*-}, \Omega^-, ...), \tag{3.1.60}$$

the wave function $\Psi_B(W)$ of the baryon corresponding to B can be expressed for example as

$$\Psi_P(W) = \langle (1,1);\, I=1/2,\, I_3=1/2,\, Y=1 | D^{(\mathbf{8})}(W) | (1,1);\, J=1/2,\, J_3,\, Y_R=1 \rangle \tag{3.1.61}$$

and

$$\Psi_{\Sigma^{*-}}(W) = \langle (3,0);\, 1, -1, 0 | D^{(\mathbf{10})}(W) | (3,0);\, 3/2,\, J_3,\, 1 \rangle. \tag{3.1.62}$$

When δH operates on the wave function, it should be represented by a linear combination of all baryon states belonging to the irreducible representation. Let us consider only the diagonal terms, ignoring the off-diagonal components as they are small. Let δH operates on a certain state and let $\langle D^{(\mathbf{8})}_{ab}(W)\rangle_B$ be the expansion coefficient of the baryon wave function in the same irreducible representation, then it is expanded as

$$D^{(\mathbf{8})}_{ab}(W)\Psi_B(W) = \langle D^{(\mathbf{8})}_{ab}(W)\rangle_{B'}\Psi_{B'}(W), \tag{3.1.63}$$

where B' runs in the octuplet. Multiplying by $\Psi^*_B(W)$ and integrating over W yields

$$\begin{aligned}\langle D^{(\mathbf{8})}_{ab}(W)\rangle_{B'} \int dW \Psi^*_B(W)\Psi_{B'}(W) &= \langle D^{(\mathbf{8})}_{ab}(W)\rangle_B \\ = \int dW \Psi^*_B(W) D^{(\mathbf{8})}_{ab}(W)\Psi_B(W),\end{aligned} \tag{3.1.64}$$

then summing the product of

$$\begin{aligned}\langle D^{(\mathbf{8})}_{ab}(W)\rangle_B &= \int dW \Psi^*_B(W) D^{(\mathbf{8})}_{ab}(W)\Psi_B(W) \\ &= \sum_\gamma \begin{pmatrix} 8 & n & n_\gamma \\ a & \mu & \mu \end{pmatrix} \begin{pmatrix} 8 & n & n_\gamma \\ b & \nu & \nu \end{pmatrix}\end{aligned} \tag{3.1.65}$$

and the Clebsch-Gordan coefficients of SU(3) over the two 8-dimensional representations gives $\langle D^{(\mathbf{8})}_{ab}(W)\rangle_B$ [18, 24]. n and n_γ are the representation dimensions of $\Psi_B(W)$, while the other quantum numbers that label $\Psi_B(W)$ are denoted $\mu = (Y, I, I_3)$ and $\nu = (Y_R, J, J_3)$.

For example, we show the quantum numbers of the baryons in the **8**-dimensional representation, where $a, b = (Y, I, I_3) = (0, 1, 0)$ or $(0, 0, 0)$ for $a, b = 3$ or 8. B is labeled by $(\mu; \nu) = (Y, I, I_3; Y_R, J, J_3)$ and $\langle D^{(\mathbf{8})}_{ab}(W)\rangle_B$ is calculated as $(y, i, i_3; y', i', i'_3) = (a; \mu)$ or $(y, i, i_3; y', i', i'_3) = (b; \nu)$ in Table 3.1. From the above discussion, the baryon mass formula becomes

Table 3.1 The quantum numbers of the 8-dimensional representation baryons

	Y	I	I_3	Y_R	J	J_3	$\langle D^{(\mathbf{8})}_{38}\rangle_B$	$\langle D^{(\mathbf{8})}_{88}\rangle_B$
P	1	1/2	1/2	1	1/2	J_3	$\frac{1}{10\sqrt{3}}$	$\frac{3}{10}$
N	1	1/2	−1/2	1	1/2	J_3	$-\frac{1}{10\sqrt{3}}$	$\frac{3}{10}$
Λ	0	0	0	1	1/2	J_3	0	$\frac{1}{10}$
Σ^+	0	1	1	1	1/2	J_3	$\frac{1}{2\sqrt{3}}$	$-\frac{1}{10}$
Σ^0	0	1	0	1	1/2	J_3	0	$-\frac{1}{10}$
Σ^-	0	1	−1	1	1/2	J_3	$-\frac{1}{2\sqrt{3}}$	$-\frac{1}{10}$
Ξ^0	−1	1/2	1/2	1	1/2	J_3	$\frac{2}{5\sqrt{3}}$	$-\frac{1}{5}$
Ξ^-	−1	1/2	−1/2	1	1/2	J_3	$-\frac{2}{5\sqrt{3}}$	$-\frac{1}{5}$

$$\begin{aligned} M = H + \delta H =& M_0 + \frac{1}{2}\Big(\frac{1}{\mathcal{I}_0} - \frac{1}{\mathcal{I}'_0}\Big)J(J+1) + \frac{1}{2}\frac{1}{\mathcal{I}'_0}\frac{1}{3}[p^2 + q^2 + pq + 3(p+q)] \\ & - \frac{1}{2}\frac{1}{\mathcal{I}'_0}\Big(\frac{\sqrt{3}}{2}\Big)^2 + \frac{1}{2}\gamma\Big(1 - \langle D^{(\mathbf{8})}_{88}(W)\rangle_B\Big) \end{aligned} \tag{3.1.66}$$

$$\mathbf{8}: \quad (p,q) = (1,1), \quad J = 1/2 \tag{3.1.67}$$

$$\mathbf{10}: \quad (p,q) = (3,0), \quad J = 3/2. \tag{3.1.68}$$

This treatment of the mass term is essentially the same for baryons in the Sakai-Sugimoto model.

3.2 Hyperons in the Sakai-Sugimoto Model

Let us analyze baryons in the SU(3) Sakai-Sugimoto model with respect to the analysis of the flavor SU(3) in the Skyrme model. The analysis will proceed as follows. First, in Sect. 3.2.1, we obtain the mass formula for the baryon. The problem of the CS term is discussed in detail in the Appendix 3.6. In Sect. 3.6.1, we summarize the problems of the CS term, and in Sects. 3.6.2 and 3.6.3, we introduce two CS terms that were proposed to solve these problems. Next, in Sect. 3.2.2, we discuss how to introduce quark masses into the Sakai-Sugimoto model.

3.2.1 Baryons in Flavor SU(3) Sakai-Sugimoto Model

As shown in (2.4.18) and (2.4.19), the action is

$$\begin{aligned} S^{DBI} =& -\frac{\lambda N_c}{108\pi^3}\int d^4xdz\mathrm{Tr}_f\Big(\frac{1}{4}K^{-1/3}F_{\mu\nu}^2+\frac{1}{2}KF_{\mu z}^2\Big) \\ &-\frac{1}{2}\kappa\int d^4xdz\Big(\frac{1}{4}K^{-1/3}\hat{F}_{\mu\nu}^2+\frac{1}{2}K\hat{F}_{\mu z}^2\Big) \end{aligned} \tag{3.2.1}$$

$$\begin{aligned} S^{CS} =& \frac{N_c}{24\pi^2}\int \omega_5^{SU(N_f)}(A)+\frac{N_c}{24\pi^2}\sqrt{\frac{2}{N_f}}\epsilon_{MNPQ}\int d^4xdz\Big[\frac{3}{8}\hat{A}_0\mathrm{Tr}_f(F_{MN}F_{PQ}) \\ &-\frac{3}{2}\hat{A}_M\mathrm{Tr}_f(\partial_0A_NF_{PQ})+\frac{3}{4}\hat{F}_{MN}\mathrm{Tr}_f(A_0F_{PQ})+\frac{1}{16}\hat{A}_0\hat{F}_{MN}\hat{F}_{PQ}-\frac{1}{4}\hat{A}_M\hat{F}_{0N}\hat{F}_{PQ} \\ &+\text{(total derivative)}\Big], \end{aligned} \tag{3.2.2}$$

for any N_f (Appendix 2.7.3). For $N_f > 2$, the significant difference from the SU(2) case is that $\omega_5(A)$ is not zero. Under the same rescale as in the SU(2) case, the DBI action becomes

$$\begin{aligned} S^{\mathrm{DBI}} =& -aN_c\int d^4xdz\mathrm{Tr}_f\Big[\frac{\lambda}{2}F_{MN}^2+\Big(-\frac{z^2}{6}F_{ij}^2+z^2F_{iz}^2-F_{0M}^2\Big)+\mathcal{O}(\lambda^{-1})\Big] \\ &-\frac{1}{2}aN_c\int d^4xdz\Big[\frac{\lambda}{2}\hat{F}_{MN}^2+\Big(-\frac{z^2}{6}\hat{F}_{ij}^2+z^2\hat{F}_{iz}^2-\hat{F}_{0M}^2\Big)+\mathcal{O}(\lambda^{-1})\Big], \end{aligned} \tag{3.2.3}$$

which is exactly the same as in the SU(2) case, where since the CS term is an integral in differential form, it does not change form under this rescale.

Classical solutions

Since the analysis is performed up to the order of λ^0, we can solve the equation of motions by considering only the highest order of $1/\lambda$ expansion. The equation of motions can be obtained as

$$\begin{aligned} A_0 :\ & D_MF_{0M}+\frac{1}{64\pi^2a}\sqrt{\frac{2}{N_f}}\epsilon_{MNPQ}\hat{F}_{MN}F_{PQ} \\ &+\frac{1}{32\sqrt{3}\pi^2a}\epsilon_{MNPQ}\Big(F_{MN}F_{PQ}-\frac{1}{N_f}\mathrm{Tr}_f(F_{MN}F_{PQ})\Big)+\mathcal{O}(\lambda^{-1})=0 \end{aligned} \tag{3.2.4}$$

$$A_M :\ D_NF_{MN}+\mathcal{O}(\lambda^{-1})=0 \tag{3.2.5}$$

$$\hat{A}_0 \ : \ \partial_M \hat{F}_{0M} + \frac{1}{64\pi^2 a}\sqrt{\frac{2}{N_f}}\epsilon_{MNPQ}\Big[\mathrm{Tr}_f(F_{MN}F_{PQ}) + \frac{1}{2}\hat{F}_{MN}\hat{F}_{PQ}\Big] + \mathcal{O}(\lambda^{-1}) \tag{3.2.6}$$

$$\hat{A}_M \ : \ \partial_N \hat{F}_{MN} + \mathcal{O}(\lambda^{-1}) = 0. \tag{3.2.7}$$

Apart from the change in coefficients due to SU(2) becoming SU(N_f), the only difference from the SU(2) case is the last term in (3.2.4). This difference is due to the fact that $\omega_5^{SU(N_f)}(A)$ is not zero, which means that the classical solution of A_0 has a finite value unlike the SU(2) case.

Let us first solve (3.2.5). The solution of this equation is obtained by embedding the BPST instanton solution (2.4.35) of SU(2) in SU(N_f) as follows;

$$A_M^{cl}(x^N) = \begin{pmatrix} A_M^{SU(2)\mathrm{BPST}} & 0 \\ 0 & 0_{N_f-2} \end{pmatrix}. \tag{3.2.8}$$

Setting this ansatz means that all the degrees of freedom of the strange quark are carried by the collective coordinates.

The equation of motions (3.2.6) and (3.2.7) can be solved exactly as in the case of SU(2) as

$$\hat{A}_0 = \frac{1}{8\pi^2 a}\frac{1}{\xi^2}\left(1 - \frac{\rho^4}{(\xi^2+\rho^2)^2}\right) \tag{3.2.9}$$

$$\hat{A}_M = 0. \tag{3.2.10}$$

Finally, let us find the solution for SU(N_f) of (3.2.4), which is the only difference from the SU(2) case. Substituting the solutions we have found so far, we get the equation

$$D_M^2 A_0 - \frac{3}{2\pi^2 a}\frac{\rho^4}{(\xi^2+\rho^2)^2}\left(\mathcal{P}_2 - \frac{2}{N_f}\mathbf{1}_{N_f}\right) = 0, \tag{3.2.11}$$

where $\mathcal{P}_2$ is

$$\mathcal{P}_2 = \begin{pmatrix} 1 & 0 & 0 \\ 0 & 1 & 0 \\ 0 & 0 & 0_{N_f-2} \end{pmatrix}, \tag{3.2.12}$$

such that it can be expressed as

$$\mathcal{P}_2 = \frac{2}{\sqrt{3}}t_8 + \frac{2}{3}\mathbf{1}_3 \tag{3.2.13}$$

in the case $N_f = 3$. This equation can be obtained as

$$A_0^{cl} = -\frac{1}{16\pi^2 a}\frac{1}{\xi^2}\Big(1 - \frac{\rho^4}{(\xi^2+\rho^2)^2}\Big)\Big(\mathcal{P}_2 - \frac{2}{N_f}\mathbf{1}_{N_f}\Big). \tag{3.2.14}$$

From the above, the mass formula for the static soliton solution is calculated to be

$$\begin{aligned}
S &= -\int dt M \\
M &= \kappa \int d^3x dz \mathrm{Tr}_f\Big[\frac{1}{2}(F_{MN}^{cl})^2 - \lambda^{-1}\Big(\frac{z^2}{6}(F_{ij}^{cl})^2 + z^2(F_{iz}^{cl})^2 - (F_{0M}^{cl})^2\Big)\Big] \\
&\quad - \frac{\kappa}{2}\lambda^{-1}\epsilon_{MNPQ}\int d^3x dz\Big[\sqrt{\frac{2}{N_f}}\frac{3}{8}\hat{A}_0^{cl}\mathrm{Tr}_f(F_{MN}^{cl}F_{PQ}^{cl}) + \frac{3}{4}\mathrm{Tr}_f(A_0^{cl}F_{MN}^{cl}F_{PQ}^{cl})\Big] \\
&\quad + \mathcal{O}(\lambda^{-1}) \\
&= 8\pi^2\kappa\Big[1 + \lambda^{-1}\Big(\frac{\rho^2}{6} + \frac{1}{320\pi^4 a^2}\frac{1}{\rho^2} + \frac{Z^2}{3}\Big)\Big].
\end{aligned} \tag{3.2.15}$$

Interestingly, this mass formula takes the same form for any N_f, despite the new contributions of $(F_{0M}^{cl})^2$ and $\mathrm{Tr}_f(A_0^{cl}F_{MN}^{cl}F_{PQ}^{cl})$ compared to SU(2) case. The minimum value of this mass formula also remains the same,

$$M_{\min} = 8\pi^2\kappa + \sqrt{\frac{2}{15}}N_c \tag{3.2.16}$$

at

$$\rho^2 = \frac{1}{8\pi^2 a}\sqrt{\frac{6}{5}}, \quad Z = 0. \tag{3.2.17}$$

The argument so far is true for arbitrary N_f. In the following, we only talk about the case $N_f = 3$.

Collective coordinates

In $N_f = 3$, the collective coordinate is

1. The position of the instanton in the $(\mathbf{x}, z)$ space : $X^M = (\mathbf{X}, Z)$
2. The size of instanton : ρ
3. SU(3) orientation : $W \in SU(3)$.

Let us consider the motion of a soliton in moduli space with time dependence on this collective coordinate. In this case, the time-dependent gauge field can be written as

$$A_M(t, x^N) = W(t) A_M^{cl}(x^N; X^\alpha(t)) W^{-1} \tag{3.2.18}$$
$$A_0(t, x^M) = W(t) A_0^{cl}(x^M; X^\alpha(t)) W(t)^{-1} + \Delta A_0(t, x^M) \tag{3.2.19}$$
$$\hat{A}_M(x^N, t) = 0 \tag{3.2.20}$$
$$\hat{A}_0(x^N, t) = \hat{A}_0^{cl}(x^M; X^\alpha(t)). \tag{3.2.21}$$

Except for A_0, the equation of motion for the static case is satisfied even if we make the collective coordinate time-dependent. For A_0, we add $\Delta A_0(t, x^M)$ because the original equation (3.2.4) (Gauss' law) cannot be satisfied if the field is time-dependent. This is a unique property to gauge theory, and although Gauss' law seems to be a redundant equation, one should always pay attention to whether it is satisfied or not (Appendix 2.6).

Let us resolve Gauss' law. First, we see that the field strength, which is time-dependent due to the insertion of the collective coordinate, is written as

$$F_{MN} = W(t) F_{MN}^{cl} W(t)^{-1} \tag{3.2.22}$$
$$F_{0M} = W(t) \Big(\dot{X}^\alpha \frac{\partial}{\partial X^\alpha} A_M^{cl} - D_M^{cl} \Phi - D_M^{cl} A_0^{cl} \Big) W(t)^{-1}, \tag{3.2.23}$$

which is defined as

$$\Phi(t, x^M) = W(t)^{-1} \Delta A_0 W(t) - i W(t)^{-1} \dot{W}(t). \tag{3.2.24}$$

Gauss' law (3.2.4) then becomes

$$D_M^{cl} \Big(\dot{X}^N \frac{\partial}{\partial X^N} A_M^{cl} + \dot{\rho} \frac{\partial}{\partial \rho} A_M^{cl} - D_M^{cl} \Phi \Big) = 0 \tag{3.2.25}$$

From this, by determining Φ, the time-dependent A_0 is solved again.

We need to consider the case where each collective coordinate is orthogonal to each other, so let's separate them like

$$\Phi = \Phi_X + \Phi_\rho + \Phi_{SU(3)} \tag{3.2.26}$$

and find the solution. For details, please refer to Appendix 3.7.1. For Φ_X and Φ_ρ, we can find

$$\Phi_X = -\dot{X}^N A_N^{cl} \tag{3.2.27}$$
$$\Phi_\rho = 0 \tag{3.2.28}$$

exactly as in SU(2). The rest can be solved for

$$D_M^{cl} D_M^{cl} \Phi_{SU(3)} = 0, \tag{3.2.29}$$

and the result is obtained as

$$\Phi_{SU(3)} = \chi^a(t)\Phi_a(x; X^\alpha(t)) \tag{3.2.30}$$

$$\Phi_a(x; X^\alpha(t)) = u^a(\xi)g(x; X(t))t_a(x; X(t))^{-1} \tag{3.2.31}$$

$$u^a(\xi) = \begin{cases} f(\xi), & (a = 1, 2, 3) \\ f(\xi)^{1/2}, & (a = 4, 5, 6, 7) \\ 1, & (a = 8) \end{cases} \tag{3.2.32}$$

$$\chi^a(t) = -2i\mathrm{tr}(t_a W(t)^{-1}\dot{W}(t)). \tag{3.2.33}$$

Since the collective coordinates give thc field time-dependent, Lagrangian L of $S_{\mathrm{YM}} + S_{\mathrm{CS}} = \int dtL$ has changed, denoting the change as δL,

$$\delta L = \delta L_{\mathrm{YM}} + \delta L_{\mathrm{CS}} \tag{3.2.34}$$

$$\begin{aligned} \delta L_{\mathrm{YM}} &= aN_c \int d^3x dz \mathrm{tr}\left(F_{0M}^2 - (F_{0M}^{cl})^2\right) \\ &= aN_c \int d^3x dz \mathrm{tr}\left(\dot{X}^N F_{MN}^{cl} + \dot{\rho}\frac{\partial}{\partial\rho}A_M^{cl} - \chi^a D_M^{cl}\Phi_a\right)^2. \end{aligned} \tag{3.2.35}$$

where δL_{YM} is the change caused by

$$F_{0M} = W(t)\left(\dot{X}^N F_{MN}^{cl} + \dot{\rho}\frac{\partial}{\partial\rho}A_M^{cl} - \chi^a D_M^{cl}\Phi^a - D_M^{cl}A_0^{cl}\right)W(t)^{-1}, \tag{3.2.36}$$

and, from this change, arises the kinetic term of each collective coordinate. δL_{CS} is zero as far as we use the CS term used in the original paper [15, 16] of the Sakai-Sugimoto model (the problems with the CS term are comprehensively explained in Appendix 3.6.1). Without considering the CS term here, the Lagrangian of the system with the collective coordinates listed above as the dynamical variables is

$$L = -M_0 + \frac{m_X}{2}\dot{\mathbf{X}}^2 + L_Z + L_\rho + L_{\rho W} + L_{CS} \tag{3.2.37}$$

$$L_Z = \frac{m_Z}{2}(\dot{Z}^2 - \omega_Z^2 Z^2) \tag{3.2.38}$$

$$L_\rho = \frac{m_\rho}{2}(\dot{\rho}^2 - \omega_\rho^2\rho^2) - \frac{K}{m_\rho\rho^2} \tag{3.2.39}$$

$$L_{\rho W} = m_\rho\rho^2\left(\frac{1}{8}\sum_{a=1}^{3}(\chi^a)^2 + \frac{1}{16}\sum_{4}^{7}(\chi^a)^2\right) \tag{3.2.40}$$

$$= 2\mathcal{I}_1(\rho)\sum_{a=3}^{3}\left[\mathrm{tr}(-iW^{-1}\dot{W}t_a)\right]^2 + 2\mathcal{I}_2(\rho)\sum_{a=4}^{7}\left[\mathrm{tr}(-iW^{-1}\dot{W}t_a)\right]^2, \tag{3.2.41}$$

where we can calculate the coefficients as follows,

$$M_0 = 8\pi^2\kappa \tag{3.2.42}$$

$$m_X = m_Z = \frac{m_\rho}{2} = 8\pi^2\kappa\lambda^{-1} = 8\pi^2 a N_c \tag{3.2.43}$$

$$\omega_Z^2 = \frac{2}{3}, \quad \omega_\rho^2 = \frac{1}{6} \tag{3.2.44}$$

$$K = \frac{N_c m_\rho}{40\pi^2 a} = \frac{2}{5}N_c^2 \tag{3.2.45}$$

$$\mathcal{I}_1(\rho) = \frac{1}{4}m_\rho\rho^2, \quad \mathcal{I}_2(\rho) = \frac{1}{8}m_\rho\rho^2. \tag{3.2.46}$$

No further analysis will yield the constraint terms necessary to obtain the baryon spectra, so we will replace the CS terms with those shown in Appendices 3.6.2 and 3.6.3, and continue the analysis by assuming that the constraint terms have arisen. In other words, we will analyze L in (3.2.37) plus

$$\frac{N_c}{2\sqrt{3}}\chi^8(t) = \frac{N_c}{\sqrt{3}}tr(-iW(t)^{-1}\dot{W}(t)t_8). \tag{3.2.47}$$

Collective coordinate quantization

The wave function can be separated as

$$\phi(z)\Psi(W)\psi(\rho), \tag{3.2.48}$$

thus the Hamiltonian can be written as

$$H = M_0 + H_Z + H_\rho + H_{\rho W} \tag{3.2.49}$$

$$H_Z = -\frac{1}{2m_Z}\partial_Z^2 + \frac{1}{2}m_Z\omega_Z^2 Z^2 \tag{3.2.50}$$

$$H_{\rho W} = \frac{1}{2\mathcal{I}_1(\rho)}\sum_{a=1}^{3}(R_a)^2 + \frac{1}{2\mathcal{I}_2(\rho)}\sum_{a=4}^{7}(R_a)^2 \tag{3.2.51}$$

$$H_\rho = -\frac{1}{2m_\rho}\frac{1}{\rho^\eta}\partial_\rho(\rho^\eta\partial_\rho) + \frac{1}{2}m_\rho\omega_\rho^2\rho^2 + \frac{K}{m_\rho\rho^2}. \tag{3.2.52}$$

For H_Z, it is the same as in the SU(2) Sakai-Sugimoto model, and for $H_{\rho W}$, the eigenvalue equations can be solved in the same way as explained in the SU(3) Skyrme model. For η of H_ρ, we can set $\eta = 8$ in the SU(3) case, but let us continue the analysis with the general η. Considering the constraint

$$R_8 = \frac{N_c}{2\sqrt{3}}, \tag{3.2.53}$$

then the Hamiltonian is

$$\begin{aligned} H_{\rho W} =& \frac{1}{2}\Big(\frac{1}{\mathcal{I}_1} - \frac{1}{\mathcal{I}_2}\Big)\sum_{b=1}^{3} R_a^2 + \frac{1}{2}\frac{1}{\mathcal{I}_2}\sum_{b=1}^{8} L_a^2 - \frac{1}{2}\frac{1}{\mathcal{I}_2} R_8^2 \\ =& \frac{2}{m_\rho \rho^2}\Big(- j(j+1) + 2\frac{1}{3}(p^2+q^2+pq+3(p+q)) - \frac{N_c^2}{6}\Big), \end{aligned} \tag{3.2.54}$$

so we can write the mass formula as

$$\begin{aligned} H\phi(z)\Psi(W)\psi(\rho) =& M_0\phi(z)\Psi(W)\psi(\rho) + (H_z\phi(z))\Psi(W)\psi(\rho) \\ &+ (H_\rho\psi(\rho))\phi(z)\Psi(W) + (H_{\rho W}\Psi(W))\phi(z)\psi(\rho) \\ =& \Big(M_0 + \sqrt{\frac{2}{3}}(n_Z + \frac{1}{2})\Big)\phi(z)\Psi(W)\psi(\rho) \\ &+ \phi(z)\Psi(W)\Big(\frac{-2j(j+1) + \frac{4}{3}(p^2+q^2+pq+3(p+q)) - \frac{N_c^2}{3}}{m_\rho\rho^2} + H_\rho\Big)\psi(\rho) \\ =& \Big(M_0 + \sqrt{\frac{2}{3}}(n_Z + \frac{1}{2}) + E^{(p,q)}_{\rho\mathrm{tot}}\Big)\phi(z)\Psi(W)\psi(\rho). \end{aligned}$$

Here, the eigenvalues of H_ρ^{tot} are $E^{(p,q)}_{\rho\mathrm{tot}}$ as follows;

$$\begin{aligned} &\Big(\frac{-2j(j+1) + \frac{4}{3}(p^2+q^2+pq+3(p+q)) - \frac{N_c^2}{3}}{m_\rho\rho^2} + H_\rho\Big)\psi(\rho) \\ &= \Big(- \frac{1}{2m_\rho}\frac{1}{\rho^\eta}\partial_\rho(\rho^\eta\partial_\rho) + \frac{1}{2}m_\rho\omega_\rho^2\rho^2 + \frac{K'}{m_\rho\rho^2}\Big)\psi(\rho) \\ &= H_\rho^{\mathrm{tot}}\psi(\rho) = E^{(p,q)}_{\rho\mathrm{tot}}\psi(\rho) \end{aligned} \tag{3.2.55}$$

$$K' = \frac{N_c^2}{15} + \frac{4}{3}(p^2+q^2+pq+3(p+q)) - 2j(j+1). \tag{3.2.56}$$

Therefore, the mass formula can be obtained by solving the eigenvalue equation

$$H_\rho^{\mathrm{tot}}\psi(\rho) = E^{(p,q)}_{\rho,\,\mathrm{tot}}\psi(\rho) \tag{3.2.57}$$

$$H_\rho^{\mathrm{tot}} = -\frac{1}{2m_\rho}\frac{1}{\rho^\eta}\partial_\rho(\rho^\eta\partial_\rho) + \frac{1}{2}m_\rho\omega_\rho^2\rho^2 + \frac{K'}{m_\rho\rho^2} \tag{3.2.58}$$

$$K' = \frac{N_c^2}{15} + \frac{4}{3}(p^2+q^2+pq+3(p+q)) - 2j(j+1) \tag{3.2.59}$$

at the end. Now, by replacing

$$\psi(\rho) = e^{-z/2} z^{\beta} v(z) \tag{3.2.60}$$

$$z = m_\rho \omega_\rho \rho^2, \quad \beta = \frac{1}{4}(\sqrt{(\eta - 1)^2 + 8K'} - (\eta - 1)), \tag{3.2.61}$$

the eigenvalue equation becomes the differential equation

$$\left(z\frac{d^2}{dz^2} + \left(2\beta + \frac{\eta+1}{2} - z\right)\frac{d}{dz} + \left(\frac{E^{(p,q)}_{\rho,\,\mathrm{tot}}}{2\omega_\rho} - \beta - \frac{\eta+1}{4}\right)\right) v(z) = 0, \tag{3.2.62}$$

which is that of the congruent hypergeometric function. Since it is known that this equation must be

$$\frac{E^{(p,q)}_{\rho,\,\mathrm{tot}}}{2\omega_\rho} - \beta - \frac{\eta+1}{4} = n, \quad n \in \mathbb{Z} \tag{3.2.63}$$

to have a regular solution, for $n = n_\rho$, we can write $v(z)$ using the confluence hypergeometric function

$$F(\alpha, \gamma; z) = \sum_{k=0}^{\infty} \frac{(\alpha)_k}{(\gamma)_k} \frac{z^k}{k!} \tag{3.2.64}$$

$$(\alpha)_k = \alpha(\alpha + 1) \cdots (\alpha + k - 1) \tag{3.2.65}$$

like

$$v(z) = F\left(-n_\rho, 2\beta + \frac{\eta+1}{2}; z\right), \tag{3.2.66}$$

whose eigenvalue is obtained as

$$E^{(p,q)}_{\rho,\,\mathrm{tot}} = \omega_\rho \left(2n_\rho + \frac{1}{2}\sqrt{(\eta - 1)^2 + 8K'} + 1\right). \tag{3.2.67}$$

From the above, the mass formula is obtained as

$$M = M_0 + \sqrt{\frac{(\eta - 1)^2}{24} + \frac{K'}{3}} + \sqrt{\frac{2}{3}}(n_\rho + n_Z + 1). \tag{3.2.68}$$

3.2.2 *Introduce Quark Masses and Obtain the Hyperon Spectra*

Several attempts have been made to introduce quark masses in the Sakai-Sugimoto model [17, 29–33]. This doctoral thesis will introduce quark masses into the Sakai-Sugimoto model according to Ref. [17] and explain the analysis of flavor SU(3) baryons in the Sakai-Sugimoto model.

Since the chiral symmetry is not explicitly broken on the original brane configuration of the Sakai-Sugimoto model, in the Ref. [17], by adding N_f D6 brane to the $D4 - D8 - \overline{D8}$ brane system, they achieve a chiral symmetry is broken. That is, we placed the D6 branes as shown in Table 3.2 and connected the $D8$-branes and $\overline{D8}$-branes. This brane configuration leads to the existence of a worldsheet instanton, and it was shown in Ref. [17] that this worldsheet instanton introduces quark masses. Furthermore, from the gravity theory side, this worldsheet instanton is found to have the form [17]

$$c \int d^4x \mathrm{Ptr}\Big[M\Big(\exp\Big[-i\int_{-z_m}^{z_m} \mathcal{A}_z dz\Big] - \mathbf{1}_{N_f}\Big)\Big] + \text{c.c.}, \tag{3.2.69}$$

which is consistent with the result obtained from the chiral perturbation theory, where c is a constant determined from experimental values as a parameter and M is the quark mass matrix.

In the following, we take into account that the action on the gravity theory side involves the deformation

$$\delta S = c \int d^4x \mathrm{Ptr}\Big[M\Big(\exp\Big[-i\int_{-z_m}^{z_m} \mathcal{A}_z dz\Big] - \mathbf{1}_{N_f}\Big)\Big] + c.c. \tag{3.2.70}$$

$$\mathrm{P}\exp\Big[-i\int_{-\infty}^{+\infty} dz \mathcal{A}_z\Big] = \exp[2i\pi(x)/f_\pi] = U(x) \tag{3.2.71}$$

$$\delta S = \int d^4x \delta L \tag{3.2.72}$$

$$\delta L = c\mathrm{tr}\big[M(U + U^\dagger - 2\mathbf{1}_{N_f})\big], \tag{3.2.73}$$

incorporating this mass term as a perturbation, as discussed in the Skyrme model. The difference from the Skyrme model is that this term depends not only on the

Table 3.2 Configuration of D6-brane

	0	1	2	3	4	5	6	7	8	9
D4	○	○	○	○	○	×	×	×	×	×
$D8-\overline{D8}$	○	○	○	○	×	○	○	○	○	○
D6	○	○	○	○	○	×	○	○	×	×

collective coordinates of the SU(3) orientation but also on ρ, so we also need to calculate the expectation value of the part that depends on ρ.

The variation δL of the Lagrangian is calculated as

$$\delta L = -\frac{4c}{3}(1 - \cos f(r))$$
$$\left[(m_u + m_d + m_s) - \frac{\sqrt{3}}{2}(m_d - m_u)D^{(8)}_{38}(G) - \frac{2m_s - m_u - m_d}{2}D^{(8)}_{88}(G)\right] \tag{3.2.74}$$

using (3.2.8) and $\hat{A}_M = 0$ of the solution obtained in Sect. 3.2.1, where we used

$$D^{(8)}_{ab}(G) = \frac{1}{2}\mathrm{tr}(G^\dagger \lambda_a G \lambda_b) \tag{3.2.75}$$

already defined in the description of the Skyrme model. Now, using the Gell-Mann-Oakes-Renner relation

$$m^2_{\pi^\pm} = \frac{2c}{f^2_\pi}(m_u + m_d), \quad m^2_{K^\pm} = \frac{2c}{f^2_\pi}(m_u + m_s), \quad m^2_{K^0,\bar{K}^0} = \frac{2c}{f^2_\pi}(m_d + m_s), \tag{3.2.76}$$

(where c is input to best reproduce the masses of $m_{\pi^\pm}$, $m_{K^\pm}$, and $m_{K^0,\bar{K}^0}$), the deformation of the mass formula due to the introduction of quark masses in the Sakai-Sugimoto model is given by

$$\delta M = -\int d^3x \delta L[A^{cl}] \tag{3.2.77}$$

$$\delta M = 4\pi f^2_\pi \rho^3 \times 1.104 \times \frac{1}{3}\big[\big(1 - \sqrt{3}D^{(8)}_{38}(G) - D^{(8)}_{88}(G)\big)m^2_{K^0,\bar{K}^0}$$
$$+ \big(1 + \sqrt{3}D^{(8)}_{38}(G) - D^{(8)}_{88}(G)\big)m^2_{K^\pm} + \big(1 + 2D^{(8)}_{88}(G)\big)m^2_{\pi^\pm}\big]. \tag{3.2.78}$$

Then, using the wave function

$$\phi(z)\Psi(W)\psi(\rho) \tag{3.2.79}$$

defined by (3.2.48), we can calculate δM to obtain the mass splitting for each baryon state. Concretely, we need to calculate

$$\int_0^\infty d\rho \rho^\eta \rho^3 \psi(\rho)^2 \tag{3.2.80}$$

and

$$\int dW \Psi_B^*(W) D_{ab}^{(\mathbf{8})}(W) \Psi_B(W), \tag{3.2.81}$$

with $\eta = 8$ for SU(3). Since calculation for the latter are described in the discussion of the Skyrme model, we will only give an explanation of the former. The wave function has the form

$$\psi(\rho) = C_{n_\rho(p,q)l} e^{-m_\rho \omega_\rho \rho^2/2} (m_\rho \omega_\rho \rho^2)^\beta F\Big(-n_\rho, 2\beta + \frac{\eta+1}{2}; m_\rho \omega_\rho \rho^2 \Big). \tag{3.2.82}$$

$C_{n_\rho(p,q)l}$ is a normalization constant, which is determined from

$$\int_0^\infty d\rho \rho^\eta \psi^2 = 1. \tag{3.2.83}$$

From the above, if we write the equation only for $n_\rho = 0$, we get

$$\langle \rho^n \rangle_{n_\rho = 0,(p,q),l} = \int_0^\infty d\rho \rho^\eta \rho^n \psi(\rho)^2 = \frac{\Gamma(2\beta + \frac{\eta+1}{2} + n/2)}{(m_\rho \omega_\rho)^{n/2} \Gamma(2\beta + \frac{\eta+1}{2})}. \tag{3.2.84}$$

This allows us to evaluate the mass splitting of baryons, but quantitatively it does not reproduce the experimental data very well (see Ref. [18]).

3.3 Dimensional Reduction

In the next section, we discuss how to introduce heavy flavor into the Sakai-Sugimoto model. We have proposed a method to implement heavy flavor in the Sakai-Sugimoto model by employing dimensional reduction while utilizing the extra-dimensional degrees of freedom. Therefore, in this section, as a first step, we explain the Forgács-Manton approach to dimensional reduction of gauge theories [34, 35].

We mentioned in Sect. 2.3.1 that the theory on the gravity side of the Sakai-Sugimoto model is a 9-dimensional YM-CS theory. In Ref. [15, 16], they have obtained five-dimensional YM-CS theory by setting the gauge field component on S^4 in the 9-dimensional spacetime $\mathbb{R}^4 \times [0, \infty) \times S^4$ to zero and considering the gauge field configurations that ignores the S^4 coordinate dependence. However, it is not necessary to treat the gauge field in this way. Therefore, it is a worthwhile attempt to review the analysis of higher-dimensional gauge theories from a more general perspective. Higher-dimensional gauge theory has been studied in Ref. [34, 36, 37] etc. (As a textbook, Ref. [38] is very helpful).

In this chapter, we will explain Forgács and Manton's argument [34, 35]. They showed that if a gauge field (G is the gauge group) is symmetric under some group S^1 as well as a metric on a space-time manifold has the same symmetry, then the YM theory can be generally reduced to a lower dimensional YM-Higgs theory.

In this thesis, we will explain the Ref. [34, 35] in the following steps. First, in the Sect. 3.3.1, we derive the symmetry equation and the consistency condition (called in this thesis following Ref. [34]), which are used to determine the form of a gauge field with a certain symmetry. It is important to understand what it means for a gauge field to be symmetric for a space-time transformation. We will then explain how to solve the symmetry equation and the consistency condition in the Sect. 3.3.2. It is easy to solve these equations on the group manifold of the group S, but from the viewpoint of dimensional reduction, it is important to be able to find solutions on the homogeneous space of this group as well. Based on the above discussion, in Sect. 3.3.3 we will discuss in general how higher dimensional gauge theories can be reduced to lower dimensional YM-Higgs theories. In the Sect. 3.3.4, we solve for the constraints that emerge in the dimensional reduction (the discussion in this section is not only for the Sakai-Sugimoto model, but can be used in general).

3.3.1 Symmetry Equations

Tensor field $T^{\rho\sigma\cdots}_{\mu\nu\ldots}$ is symmetric (Appendix 3.5) with respect to the group S generated by the generator ξ_m means that

$$\mathbf{L}_{\xi_m} T^{\rho\sigma\ldots}_{\mu\nu\ldots} = 0 \tag{3.3.1}$$

is satisfied by the Lie derivative $\mathbf{L}_{\xi_m}$ (see also Fig. 2.4), where

$$\begin{aligned}\mathbf{L}_{\xi_m} T^{\rho\sigma\ldots}_{\mu\nu\ldots} &= (\partial_\mu \xi^\lambda) T^{\rho\sigma\ldots}_{\lambda\nu\ldots} + (\partial_\nu \xi^\lambda) T^{\rho\sigma\ldots}_{\mu\lambda\ldots} + \cdots \\ &- (\partial_\lambda \xi^\rho) T^{\lambda\sigma\ldots}_{\mu\nu\ldots} - (\partial_\lambda \xi^\sigma) T^{\rho\lambda\ldots}_{\mu\lambda\ldots} - \cdots \quad . \\ &+ \xi^{\lambda\partial}_\lambda T^{\rho\sigma\ldots}_{\mu\nu\ldots}\end{aligned} \tag{3.3.2}$$

On the other hand, a gauge field is symmetric under a group S if the gauge transformation absorbs the change caused by the symmetric transformation (see Fig. 3.2).

We write $g = \exp(W(x))$ for the elements of G with $W(x) = W^a(x)T^a$ (T^a is the generator of G) and expand it to $W(x) = \lambda_m W_m(x)$ with the basis λ_m of the vector space of $\xi = \lambda_m \xi_m$. Then, (3.3.1) is the weakened condition

$$\mathbf{L}_{\xi_m} A_\mu = D_\mu W_m, \tag{3.3.3}$$

[1] The symmetry of vector fields was outlined in Fig. 2.4, but in the case of gauge fields, the restriction is weakened, as will be explained later.

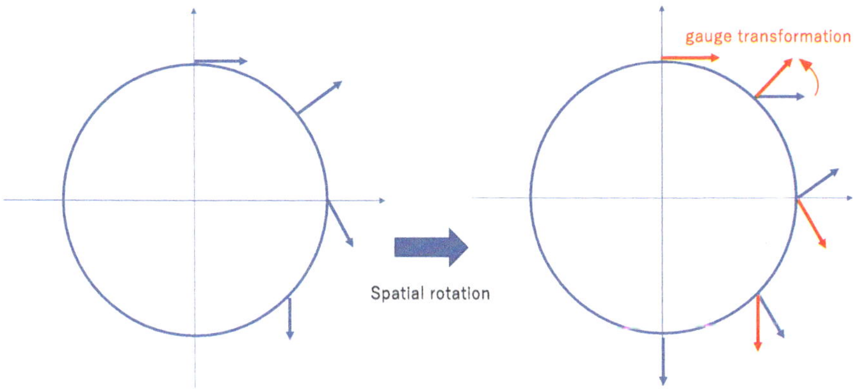

Fig. 3.2 Symmetric gauge fields

where W_m is the Lie algebra of the gauge group G and $D_\mu W_m = \partial_\mu W_m + [A_\mu, W_m]$. Now we found that the symmetry equations of the gauge field A_μ are

$$(\partial_\mu \xi_m^\rho) A_\rho + \xi_m^\rho (\partial_\rho A_\mu) = D_\mu W_m, \tag{3.3.4}$$

where m is the number of generators of the group S. W transforms as a scalar for the coordinate transformations, and its gauge transformation is

$$W^g = gWg^{-1} + \xi^\rho(\partial_\rho g)g^{-1} \tag{3.3.5}$$

$$= gWg^{-1} + (\mathbf{L}_\xi g)g^{-1}. \tag{3.3.6}$$

This means that the symmetry equation is an equation that depends neither on the choice of coordinates nor on the choice of gauge.

We will show the definition of the symmetry generator ξ_m and discuss how ξ_m can be obtained. It is important to note that ξ is defined not only on S, but also on coset space S/R. The generator ξ_m of a group S (of dimension N) is a vector field such that it satisfies

$$[\xi_m, \xi_n]^\mu = f_{mnp}\xi_p^\mu \tag{3.3.7}$$

using Lie brackets

$$\eta^\mu = [\xi_m, \xi_n]^\mu = \xi_m^\rho \partial_\rho \xi_n^\mu - \xi_n^\rho \partial_\rho \xi_m^\mu.$$

Since the Lie bracket operator is antisymmetric to satisfy the Jacobi identity, it is a representation of the Lie algebra of the group S. We now turn from the discussion on space-time manifolds to that on group manifolds. When we use μ in the subscripts, it means that we have a space-time manifold in mind; otherwise, we use α or $\hat{\alpha}$.

The relation between arguments on group manifolds and space-time manifolds will be discussed in Sect. 3.3.3. The most fundamental solution of (3.3.7) is the set of infinitesimal right transformations of the group S. Suppose that the group manifold S has coordinate $y^{\hat{\alpha}}$. Then, we define the infinitesimal right transformation $\xi^{\hat{\alpha}}(y^{\hat{\beta}})$ to be

$$s \times (1 + \epsilon J^m) = s(y^{\hat{\alpha}} + \epsilon \xi_m^{\hat{\alpha}}) \simeq s + \epsilon \xi_m^{\hat{\alpha}} \partial_{\hat{\alpha}} s, \tag{3.3.8}$$

where $s(y^{\hat{\alpha}}) \in S$. It means that the result of the group operator on the left-hand side causes a transformation of the coordinates as shown on the right-hand side. This equation can be written as

$$s J^m = \mathbf{L}_{\xi_m} s. \tag{3.3.9}$$

From this equation, we can obtain $\xi_m^{\hat{\alpha}}$. We can verify that the infinitesimal right transform $\xi^{\hat{\alpha}}(y^{\hat{\beta}})$ defined in this way satisfies (3.3.7) as follows. The action of the commutator of the Lie derivative on s is written as

$$(\mathbf{L}_{\xi_m} \mathbf{L}_{\xi_n} - \mathbf{L}_{\xi_n} \mathbf{L}_{\xi_m}) s = s[J^m, J^n] = f_{mnp} s J^p \tag{3.3.10}$$

This expression is transformed into

$$\mathbf{L}_{[\xi_m, \xi_n]} s = f_{mnp} \mathbf{L}_{\xi_p} s \tag{3.3.11}$$

(Appendix 3.7.2), from which we can derive (3.3.7).

We can also obtain a solution of (3.3.7) from a right transformation on the right coset space of S. Let R be a subgroup of S of dimension $N - N'$ such that it has generator J^m, $(N' + 1 \leqq m \leqq N)$, Then this coset Rs has N' coordinates y^α and

$$Rs \times (1 + \epsilon J^m) = Rs + \epsilon \xi_m^\alpha \partial_\alpha Rs \tag{3.3.12}$$

is obtained by the same argument as before.

We can assign y^α to the coset space S/R and y^ω to the subgroup R. If we determine the representative element of coset S/R to be

$$s_0(y^\alpha) \in Rs(y^\alpha) \tag{3.3.13}$$

in each class, all elements of S can be uniquely written as

$$s(y^{\hat{\alpha}}) = r(y^\omega) s_0(y^\alpha) \tag{3.3.14}$$

by $y^{\hat{\alpha}} = (y^\omega, y^\alpha)$.

Under the transformation of the group S, we can find the relation satisfied by the Lie algebra W_m used to absorb the change in A_μ. The commutator

$$(\mathbf{L}_{\xi_m}\mathbf{L}_{\xi_n} - \mathbf{L}_{\xi_n}\mathbf{L}_{\xi_m})A_\mu = \mathbf{L}_{\xi_m}(D_\mu W_n) - \mathbf{L}_{\xi_n}(D_\mu W_m)$$

of the Lie derivative becomes

$$\mathbf{L}_{f_{mnp}\xi_p} A_\mu = D_\mu(\mathbf{L}_{\xi_m} W_n - \mathbf{L}_{\xi_n} W_m + [W_m, W_n]), \tag{3.3.15}$$

with the left and right sides transformed, respectively (Appendix 3.7.3). From this, we obtain

$$\mathbf{L}_{\xi_m} W_n - \mathbf{L}_{\xi_n} W_m + [W_m, W_n] - f_{mnp} W_p = 0. \tag{3.3.16}$$

The equation that W_m must satisfy is called the consistency condition, following Ref. [34]. Therefore, when there is a pair of Lie algebras of G such that (3.3.16) is satisfied, we can obtain an S-symmetric gauge field A_μ by applying the constraints of (3.3.4).

In the rest of this section, we will find some relations that we will use later. From (3.3.4), the symmetry equation for $F_{\mu\nu} = \partial_\mu A_\nu - \partial_\nu A_\mu + [A_\mu, A_\nu]$ is given by

$$\begin{aligned} \mathbf{L}_{\xi_m} F_{\mu\nu} &= D_\mu \mathbf{L}_{\xi_m} A_\nu - D_\nu \mathbf{L}_{\xi_m} A_\mu = (D_\mu D_\nu - D_\nu D_\mu) W_m \\ &= -[F_{\mu\nu}.W] \end{aligned} \tag{3.3.17}$$

(Appendix 3.7.4). Using (3.3.4), we obtain

$$\xi_m^\mu F_{\mu\nu} = D_\nu(W_m - \xi_m^\mu A_\mu) \tag{3.3.18}$$

(Appendix 3.7.5). If we define $\Psi_m = \xi_m^\mu A_\mu - W_m$, we can write this as

$$\xi_m^\mu F_{\mu\nu} = -D_\nu \Psi_m. \tag{3.3.19}$$

Furthermore, using (3.3.18), (3.3.16), and (3.3.7), we obtain

$$\xi_m^\mu \xi_n^\nu F_{\mu\nu} = f_{mnp}\Psi_p + [\Psi_m, \Psi_n] \tag{3.3.20}$$

(Appendix 3.7.6). The Ψ_m is related to the Higgs sector after dimensional reduction.

3.3.2 The Solution of the Symmetry Equation

Again, symmetry equations and consistency conditions are given as

$$\mathbf{L}_{\xi_m} A_\mu = D_\mu W_m \tag{3.3.21}$$

$$\mathbf{L}_{\xi_m} W_n - \mathbf{L}_{\xi_n} W_m + [W_m, W_n] - f_{mnp} W_p = 0. \tag{3.3.22}$$

Since both symmetry generators (3.3.8) and (3.3.12) for S and S/R satisfy the commutation relation (3.3.7), solutions for (3.3.21) and (3.3.22) exist for both S and S/R. From the viewpoint of dimensional reduction, the solution defined on S/R is important. The consistency condition on S/R is

$$\mathbf{L}_{\xi_m} W_n(y^\alpha) - \mathbf{L}_{\xi_n} W_m(y^\alpha) + [W_m(y^\alpha), W_n(y^\alpha)] - f_{mnp} W_p(y^\alpha) = 0. \quad (3.3.23)$$

Now, W_m is redefined as

$$W_m(y^\omega, y^\alpha) = W_m(y^\alpha) \quad \forall y^\alpha. \quad (3.3.24)$$

For W_m, it is sufficient that exist the ones satisfying (3.3.21) and (3.3.22), so such a redefinition is no problem as long as it can be solved. It is easy to conclude that this W_m also satisfies the consistency condition on S. That is,

$$\mathbf{L}_{\xi_m} W_n(y^{\hat{\alpha}}) - \mathbf{L}_{\xi_n} W_m(y^{\hat{\alpha}}) + [W_m(y^{\hat{\alpha}}), W_n(y^{\hat{\alpha}})] - f_{mnp} W_p(y^{\hat{\alpha}}) = 0 \quad (3.3.25)$$

is satisfied. If we define the new field $W_{\hat{\alpha}}$ as

$$W_m = \xi_m^{\hat{\alpha}} W_{\hat{\alpha}}, \quad (3.3.26)$$

Equation (3.3.25) becomes

$$\partial_{\hat{\alpha}} W_{\hat{\beta}} - \partial_{\hat{\beta}} W_{\hat{\alpha}} + [W_{\hat{\alpha}}, W_{\hat{\beta}}] = 0 \quad (3.3.27)$$

(Appendix 3.7.7), and thus $W_{\hat{\alpha}}$ can be expressed as

$$W_{\hat{\alpha}} = (\partial_{\hat{\alpha}} g) g^{-1} \quad (3.3.28)$$

in pure gauge on S. Although W_m satisfies the consistency condition both on S and on S/R, by considering the fields on S, W_m can also vanish because $W_{\hat{\alpha}}$ can be set to zero by gauge transformation using (3.3.5).[2]

Symmetry equation is also considered in the same way as the consistency condition. The symmetry equation on S/R is

$$\begin{aligned} \mathbf{L}_{\xi_m} A_\alpha &= D_\alpha W_m \\ \Leftrightarrow (\partial_\alpha \xi_m^\beta) A_\beta + \xi^\beta \partial_\beta A_\alpha &= \partial_\alpha W_m - [A_\alpha, W_m] \end{aligned} \quad (3.3.29)$$

using $A_\alpha(y^\beta)$ and $W_m(y^\alpha)$. In order to embed in S, $W_m(y^\alpha)$ and $A_\alpha(y^\beta)$ are redefined by (3.3.24) and

$$A_{\hat{\alpha}}(y^{\hat{\beta}}) = (A_\omega(y^{\hat{\beta}}), A_\alpha(y^{\hat{\beta}})) = (0, A_\alpha(y^\beta)), \quad (3.3.30)$$

[2] If we consider on S/R, we can only perform gauge transformations independent of y^ω.

respectively. This restriction on $A_{\hat{\alpha}}(y^{\hat{\beta}})$ should be understood as the restriction that symmetry equations impose on gauge fields. This restriction will be discussed later. In this case, because of

$$(\partial_\alpha \xi_m^\beta) A_\beta = (\partial_\alpha \xi_m^{\hat{\beta}}) A_{\hat{\beta}} \tag{3.3.31}$$

$$\xi^\beta \partial_\beta A_\alpha = \xi^{\hat{\beta}} \partial_{\hat{\beta}} A_\alpha, \tag{3.3.32}$$

Equation (3.3.29) can be written as

$$(\partial_\alpha \xi_m^{\hat{\beta}}) A_{\hat{\beta}} + \xi^{\hat{\beta}} \partial_{\hat{\beta}} A_\alpha = \partial_\alpha W_m - [A_\alpha, W_m]. \tag{3.3.33}$$

On the other hand, considering that ξ_m^α is independent of y^ω, (3.3.24), and (3.3.30), then

$$(\partial_\omega \xi_m^{\hat{\beta}}) A_{\hat{\beta}} + \xi^{\hat{\beta}} \partial_{\hat{\beta}} A_\omega = \partial_\omega W_m - [A_\omega, W_m] \tag{3.3.34}$$

automatically holds. Therefore, by combining these two equations, we obtain the symmetry equation on S,

$$\mathbf{L}_{\xi_m} A_{\hat{\alpha}} = D_{\hat{\alpha}} W_m, \tag{3.3.35}$$

with $A_\alpha(y^\beta)$ and $W_m(y^\alpha)$ defined on S/R. As mentioned above, since W_m can be set to zero by embedding on S, the symmetry equation can be simplified to the form

$$\mathbf{L}_{\xi_m} A_{\hat{\alpha}} = 0, \tag{3.3.36}$$

which is known how to solve. Now the restriction to $A_{\hat{\alpha}}$ is attributed to (3.3.30) and (3.3.36).

Instead of solving (3.3.36), we solve the more general equation

$$\mathbf{L}_{\xi_m} T^{\hat{\gamma}\hat{\delta}\dots}_{\hat{\alpha}\hat{\beta}\dots} = 0. \tag{3.3.37}$$

We defined the symmetry generator (vector field) ξ_m as an infinitesimal right transformation in (3.3.8). In the same way, the infinitesimal left transformation $\tilde{\xi}_m$ can be defined as

$$(1 + \epsilon J^m) s = s - \epsilon \xi_m^{\hat{\alpha}} \partial_{\hat{\alpha}} s \tag{3.3.38}$$

$$\Leftrightarrow J^m s = -\mathbf{L}_{\xi_m} s. \tag{3.3.39}$$

The vector field $\tilde{\xi}_m \xi_m$ satisfies Lie bracket algebra as well as ξ_m. In fact, this infinitesimal left transformation turns out to be

$$\mathbf{L}_{\xi_m}\tilde{\xi}_n^{\hat{\alpha}} = 0. \tag{3.3.40}$$

This is trivial from the fact that the Lie derivative and Lie bracket are equivalent with

$$\mathbf{L}_{\tilde{\xi}_n}\mathbf{L}_{\xi_m}s = -J^n(sJ^m) = -(J^n s)J^m = \mathbf{L}_{\xi_m}\mathbf{L}_{\tilde{\xi}_n}s$$
$$\Leftrightarrow [\xi_m, \tilde{\xi}_n] = 0 \tag{3.3.41}$$

in the present case, by noting that $\tilde{\xi}_n^{\hat{\alpha}}$ is a tensor with superscripts. Furthermore, if we define as the inverse $\tilde{\xi}_{m\hat{\alpha}}$,

$$\tilde{\xi}_{m\hat{\alpha}}\tilde{\xi}_n^{\hat{\alpha}} = \delta_{mn}, \quad \tilde{\xi}_{m\hat{\alpha}}\tilde{\xi}_m^{\hat{\beta}} = \delta_{\hat{\alpha}}^{\hat{\beta}} \tag{3.3.42}$$

of $\tilde{\xi}_m^{\hat{\alpha}}$ with $(m, \hat{\alpha})$, it is easy to show that this vector field also satisfies

$$\mathbf{L}_{\xi_m}\tilde{\xi}_{m\hat{\alpha}} = 0 \tag{3.3.43}$$

by Leibniz rule. Based on the above discussion, the solution of (3.3.37) is given by

$$T^{\hat{\gamma}\hat{\delta}....}_{\hat{\alpha}\hat{\beta}....} = \lambda_{mn...pq...}\tilde{\xi}_{m\hat{\alpha}}\tilde{\xi}_{n\hat{\beta}}...\tilde{\xi}_p^{\hat{\gamma}}\tilde{\xi}_q^{\hat{\delta}} \tag{3.3.44}$$

with the constant $\lambda_{mn...pq...}$.

Now the solution of symmetry equation (3.3.36) can be expressed as

$$A^a_{\hat{\alpha}} = \Phi^a_m(x)\tilde{\xi}_{m\hat{\alpha}}(y), \tag{3.3.45}$$

with $\lambda_m = \Phi^a_m(x)$. Since Φ only has not to depend on y, we have given it an x dependence. In the discussion so far, we have considered only group manifolds, so we do not need to consider x-dependence. However, as we will discuss later, when the group S acts on the space-time manifold, we introduced the x-dependence here because it exists as a coordinate of a space not related to the group S. The reason A_ω components seem to remain is that we have performed a gauge transformation to set W_m to zero. Furthermore, setting $\lambda_{mn} = \delta_{mn}$ gives the rank two tensor,

$$\tilde{h}^{\hat{\alpha}\hat{\beta}} = \tilde{\xi}_m^{\hat{\alpha}}\tilde{\xi}_m^{\hat{\beta}}. \tag{3.3.46}$$

This is the (left) metric tensor of the Lie group, since it gives the inner product of two vectors, like

$$\tilde{\xi}_{m\hat{\alpha}}\tilde{h}^{\hat{\alpha}\hat{\beta}} = \tilde{\xi}_n^{\hat{\beta}}. \tag{3.3.47}$$

We can also define the (right) metric tensor $h^{\hat{\alpha}\hat{\beta}} = \xi_m^{\hat{\alpha}}\xi_m^{\hat{\beta}}$, which now satisfies

$$\begin{aligned}\mathbf{L}_{\xi_m} h^{\hat{\alpha}\hat{\beta}} &= (\mathbf{L}_{\xi_m}\xi_n)^{\hat{\alpha}}\xi^{\hat{\beta}} + \xi_n^{\hat{\alpha}}(\mathbf{L}_{\xi_m}\xi_n)^{\hat{\beta}} \\ &= f_{mnp}\xi_p^{\hat{\alpha}}\xi_n^{\hat{\beta}} + f_{mnp}\xi_n^{\hat{\alpha}}\xi_p^{\hat{\beta}} \\ &= 0,\end{aligned} \tag{3.3.48}$$

because the Lie derivative and the Lie brackets are equivalent. This means that $h^{\hat{\alpha}\hat{\beta}}$ and $\tilde{h}^{\hat{\alpha}\hat{\beta}}$ are covariant for symmetric transformations that shift with $\xi_m^{\hat{\alpha}}$. Moreover, with

$$\begin{aligned}&\mathbf{L}_{\xi_m} e = \xi_m^{\hat{\alpha}}\partial_{\hat{\alpha}} e = eJ^m = J^m \\ &\mathbf{L}_{\tilde{\xi}_m} e = \tilde{\xi}_m^{\hat{\alpha}}\partial_{\hat{\alpha}} e = -J^m e = -J^m \\ &\Leftrightarrow \xi_m^{\hat{\alpha}} = -\tilde{\xi}_m^{\hat{\alpha}},\end{aligned} \tag{3.3.49}$$

we conclude

$$h^{\hat{\alpha}\hat{\beta}} = \tilde{h}^{\hat{\alpha}\hat{\beta}}, \tag{3.3.50}$$

which holds for arbitrary elements, because the metric keeps on the Lie group in the present case. We also note that a similar argument yields

$$f_{mnp}\tilde{\xi}_m^{\hat{\alpha}}\tilde{\xi}_n^{\hat{\beta}}\tilde{\xi}_p^{\hat{\gamma}} = -f_{mnp}\xi_m^{\hat{\alpha}}\xi_n^{\hat{\beta}}\xi_p^{\hat{\gamma}}, \tag{3.3.51}$$

which we will need later.

Finally, we discuss the constraint (3.3.30) imposed on $A_{\hat{\alpha}}(y^{\hat{\beta}})$ when the symmetry equation is embedded. This requires that $A_\omega = 0$ and that the remaining components are independent of y^ω. A necessary and sufficient condition for this constraint is that $F_{i\omega},\ F_{\alpha\omega},\ F_{\omega\tau}$ must all be zero. $F_{\omega\tau} = 0$ requires $A_\omega = 0$ while $F_{i\alpha} = 0,\ F_{\alpha\omega} = 0$ requires $A_i,\ A_\alpha$ to be independent of y^ω respectively. Noting that we can now set $W_m = 0$, from (3.3.18) and (3.3.20), the field strength are written by

$$F_{\hat{\alpha}i} = \tilde{\xi}_{m\hat{\alpha}}(\partial_i\Phi_m + [A_i, \Phi_m]) \tag{3.3.52}$$

$$F_{\hat{\alpha}\hat{\beta}} = -\tilde{\xi}_{m\hat{\alpha}}\tilde{\xi}_{n\hat{\beta}}(f_{mnp}\Phi_p + [\Phi_m, \Phi_n]) \tag{3.3.53}$$

(we used (3.3.51) when obtaining the second equation), so in order to set $F_{i\omega},\ F_{\hat{\alpha}\omega}$ to zero, we require

$$\tilde{\xi}_{m\alpha}(\partial_i\Phi_m + [A_i, \Phi_m]) = 0 \tag{3.3.54}$$

$$\tilde{\xi}_{n\omega}(f_{mnp}\Phi_p + [\Phi_m, \Phi_n]) = 0. \tag{3.3.55}$$

The meaning of the above equation can be understood as follows. For $m > N'$, $\tilde{\xi}_m^\omega$ is also the symmetry generator of the subgroup R, so the following equation obtains

$$\tilde{\xi}_m^{\hat{\alpha}} \partial_{\hat{\alpha}} s = -J^m s \tag{3.3.56}$$

$$\tilde{\xi}_m^{\omega} \partial_{\omega} s = -J^m s, \tag{3.3.57}$$

from which conclude

$$\tilde{\xi}_m^{\alpha} = 0, \quad m > N'. \tag{3.3.58}$$

Then,

$$\tilde{\xi}_{m\omega} = 0, \quad m \leq N' \tag{3.3.59}$$

is also trivially satisfied. In the case $m > N'$ (where N' is determined to be the Lie algebra of R), $\tilde{\xi}_{m\omega}$ is an infinitesimal left transformation of R. On the other hand, in the case $m \leq N'$, $\tilde{\xi}_{m\omega}$ is zero. Therefore, (3.3.54) and (3.3.54) become

$$\partial_i \Phi_n + [A_i, \Phi_n] = 0 \quad n > N' \tag{3.3.60}$$

$$f_{mnp} \Phi_p + [\Phi_m, \Phi_n] = 0 \quad n \leq N', \tag{3.3.61}$$

which is the final form of the constraint we wanted (Appendices 3.7.8 and 3.7.8).

Taking this constraint into account, we can further discuss about (3.3.45). First, from (3.3.61), by setting $\Phi'_m = -\Phi_m$ ($m > N'$), Φ'_m can be regarded as the generator of the subgroup R of G. Then Φ'_m is a constant and (3.3.60) is

$$[A_i, \Phi_m] = 0. \tag{3.3.62}$$

Here, looking at (3.3.45), A_ω are non-zero. As already mentioned, one of the implications of constraint was that this component is zero. By gauge transformation, we can show that this A_ω component can be set to zero. For $m > N'$, J^m and $-\Phi_m$ are generators of S and the subgroup R of G, respectively, so A_ω is written by

$$A_\omega = \Phi_m \tilde{\xi}_{m\omega} = -J^m \tilde{\xi}_{m\omega} = (\partial_\omega r) r^{-1} \tag{3.3.63}$$

(Appendix 3.7.10), where $r(y^\omega) = \exp(\alpha_m(y^\omega) J^m) = \exp(\alpha_m(y^\omega)(-\Phi_m))$ and we used

$$(\partial_\omega r) r^{-1} = -J^m \tilde{\xi}_{m\omega}. \tag{3.3.64}$$

Since r is also an element of the gauge group, we can vanish as follows;

$$\begin{aligned} A_\omega \rightarrow & r^{-1} A_\omega r + (\partial_\omega r^{-1}) r = r^{-1} (\partial_\omega r) r^{-1} r + (\partial_\omega r^{-1}) r \\ & = r^{-1} (\partial_\omega r) + (\partial_\omega r^{-1}) r = 0. \end{aligned} \tag{3.3.65}$$

Since r commutes with A_i, the other gauge field components are

$$A_i = A_i(x) \tag{3.3.66}$$

$$A_\alpha = r^{-1}\Phi_m r\tilde{\xi}_{m\alpha} \tag{3.3.67}$$

under this gauge transformation, where i is the index that labels the space-time manifold that is not affected by S transformations. Note that A_i is $y^{\hat{\alpha}}$ independent because this fields should also be S-symmetric, which means

$$\begin{aligned}\mathbf{L}_{\xi_m} A_i(x, y) &= (\partial_i \xi^\mu) A_\mu + \xi^\mu(\partial_\mu A_i(x, y)) \\ &= \xi^\alpha(\partial_\alpha A_i(x, y)) = 0,\end{aligned} \tag{3.3.68}$$

with $\xi^\mu(x, y) = (\xi^i(x, y), \xi^\alpha(x, y)) = (0, \xi^\alpha(y^\beta))$. Under this gauge fixing, $W_{\hat{\alpha}}$ is

$$W_\omega = -r^{-1}(\partial_\omega r) = -r^{-1}(\partial_\omega r)r^{-1}r = -r^{-1}\Phi_m\tilde{\xi}_{m\omega}r \tag{3.3.69}$$

$$W_\alpha = 0 \tag{3.3.70}$$

so $W_m = \xi^{\hat{\alpha}}_m W_{\hat{\alpha}}$ is

$$W_m = -r^{-1}\Phi_n r\xi^\omega_m\tilde{\xi}_{m\omega}, \tag{3.3.71}$$

which allows A_ω to be absorbed within W_m. Since A_α and W_m are independent of y^ω, we can always take $r = 1$ by choosing r appropriately. Thus, it is simplified to

$$A_\alpha = \Phi_m\tilde{\xi}_{m\alpha} \tag{3.3.72}$$

$$W_m = -\Phi_n\xi^\omega_m\tilde{\xi}_{n\omega}. \tag{3.3.73}$$

3.3.3 Dimensional Reduction of the Action

When the subspace of a space-time manifold is $\mathscr{X}$ and the group S acts transitively on $\mathscr{X}$, we have a homogeneous space $S/R \cong \mathscr{X}$. Then the metric of $\mathscr{X}$ is proportional to the metric $h^{\alpha\beta}$ on S/R. The S-symmetric gauge fields that lived on the group manifold S discussed in the previous section are also S-symmetric gauge fields on $\mathscr{X}$. For example, $\mathbb{R}^3$ and S^3 can be divided into $[0, \infty) \times S^2$. Since SO(3) acts transitively on S^2 (corresponding to $\mathscr{X}$), SO(3)/SO(2) $\cong S^2$. When, for example, the metric of a space-time manifold can be written as

$$h^{\mu\nu}(x^i, y^\alpha) = \left(\begin{array}{c|c} h^{ij}(x) & 0 \\ \hline 0 & \frac{1}{R^2(x)}h^{\alpha\beta}(y) \end{array}\right), \tag{3.3.74}$$

dimensional reduction is possible in general.

If the original action can now be written as

$$\mathscr{L} = -\frac{1}{8}Tr\int d^D x^\mu h^{1/2} F_{\mu\nu} F_{\sigma\tau} h^{\mu\sigma} h^{\nu\tau}, \tag{3.3.75}$$

it can be expanded to

$$\mathscr{L} = -\frac{1}{8}Tr\int d^{D'}x^i d^{N'}y^\alpha h^{1/2}\Big[F_{ij}F_{kl}h^{ik}h^{jl} + \frac{2}{R^2}F_{i\alpha}F_{j\beta}h^{ij}h^{\alpha\beta} + \frac{1}{R^4}F_{\alpha\beta}F_{\gamma\delta}h^{\alpha\gamma}h^{\beta\delta}\Big]. \tag{3.3.76}$$

If we gauge transform $A_\omega = 0$ and note that $h^{\hat{\alpha}\hat{\beta}} = \tilde{h}^{\hat{\alpha}\hat{\beta}}$, the second term of this action can be calculated as

$$\begin{aligned}
F_{i\alpha}F_{j\beta}h^{ij}h^{\alpha\beta} &= (\partial_i A_\alpha - \partial_\alpha A_i - [A_i, A_\alpha])(\partial_j A_\beta - \partial_\beta A_j - [A_j, A_\beta])\tilde{h}^{\alpha\beta}h^{ij} \\
&= (\partial_i r^{-1}\Phi_m r - [A_i, r^{-1}\Phi_m r])\tilde{\xi}_{m\alpha}(\partial_j r^{-1}\Phi_n r - [A_j, r^{-1}\Phi_n r])\tilde{\xi}_{n\beta}\tilde{h}^{\alpha\beta}h^{ij} \\
&= r^{-1}(\partial_i\Phi_m - [A_i, \Phi_m])\tilde{\xi}_{m\hat{\alpha}}(\partial_j\Phi_n - [A_j, \Phi_n])\tilde{\xi}_{n\hat{\beta}}\tilde{\xi}_k^{\hat{\alpha}}\tilde{\xi}_k^{\hat{\beta}}h^{ij}r \\
&= r^{-1}D_i\Phi_k D_j\Phi_k h^{ij}r,
\end{aligned} \tag{3.3.77}$$

which is no longer y^ω dependent. We also see that r disappears from the action without taking $r = 1$. This result can also be obtained using (3.3.19). The third term can be calculated in the same way as

$$\begin{aligned}
F_{\alpha\beta}F_{\gamma\delta}h^{\alpha\gamma}h^{\beta\delta} &= (f_{rst}\Phi_t + [\Phi_r, \Phi_s])(f_{rst}\Phi_t + [\Phi_r, \Phi_s]) \\
&= 2V(\Phi).
\end{aligned} \tag{3.3.78}$$

From the above, the action is reduced to

$$\mathscr{L} = \Omega\int d^{D'}x R^{N'}(deth_{ij})^{1/2}Tr\Big[\frac{1}{4}F_{ij}F_{kl}h^{ik}h^{jl} + \frac{1}{2R^2}D_i\Phi_m D_j\Phi_m h^{ij} + \frac{1}{2R^4}V(\Phi)\Big] \tag{3.3.79}$$

by dimension reduction.

3.3.4 Solution of Constraint

A general method for solving the constraints arising from dimensional reduction is given by Manton in Ref. [35]. The constraint we want to solve now is for the SU(2+1) case including c and b quarks, which is also specifically calculated for the

SU(3) gauge group in this paper. In this case, the constraints (3.3.60) and (3.3.61) we want to solve now become

$$\partial_\mu \Phi_3 + [A_\mu, \Phi_3] = 0 \tag{3.3.80}$$
$$\partial_z \Phi_3 + [A_z, \Phi_3] = 0 \tag{3.3.81}$$
$$f_{m3p}\Phi_p + [\Phi_m, \Phi_3] = 0. \tag{3.3.82}$$

First, let us briefly review the essential knowledge of group theory required in this section. For details, please refer to Ref. [35]. In this section, we will use a Chevalley basis, which allows us to relate the properties of commutation relations and Killing forms to root usefully.

Choose all d bases of any compact Lie algebra $\mathfrak{g}$ that are commutative with each other and denote them by H_i, then H_i satisfying

$$[H_i, H_j] = 0, \quad (i, j = 1, ..., \text{rank}(\mathfrak{g})) \tag{3.3.83}$$

is called a Cartan subalgebra and its maximum number is denoted by rank $(\mathfrak{g})$. The adjoint representation of all H_i has a common eigenvector v_ω of $d - \text{rank}(\mathfrak{g})$ eigenvalues ω_i (ω denotes each root). For a v_ω, ω_i is called the root of $\mathfrak{g}$. The root is a finite set of vectors spanning rank$(\mathfrak{g})$-dimensional Euclidean space. Each root has a one-to-one correspondence with its generator, which we denote by χ_ω. In this case, we can replace the basis of $\mathfrak{g}$ with a new d basis of H_i and χ_ω. The commutation relation of the newly chosen generators, including (3.3.83), is

$$[H_i, \chi_\omega] = \omega_i \chi_\omega \tag{3.3.84}$$
$$[\chi_\omega, \chi_{-\omega}] = \frac{2\omega_i}{\omega \cdot \omega} \tag{3.3.85}$$
$$[\chi_\omega, \chi_\tau] = (r+1)\chi_{\omega+\tau}. \tag{3.3.86}$$

Furthermore, the Killing form has the property

$$\begin{aligned}
&(H_i, \chi_\omega) = 0\\
&(\chi_\omega, \chi_{-\omega}) = \frac{2}{\omega \cdot \omega}\\
&(\chi_\omega, \chi_\tau) = 0, \quad (\tau \neq \omega).
\end{aligned} \tag{3.3.87}$$

An ω string through τ means a set of roots $\tau - r\omega, ..., \tau + q\omega$ consisting of root ω and τ. The r, q is the largest integer for which $\tau - r\omega, ..., \tau + q\omega$. If we define the quantity

$$\langle \tau, \omega \rangle = \frac{2\tau \cdot \omega}{\omega \cdot \omega}, \tag{3.3.88}$$

then

$$r - q = \langle \tau, \omega \rangle. \tag{3.3.89}$$

holds.

In particular, in the case of SU(3), we consider a γ string through α consisting of two roots of $\alpha - \gamma$, α because we will use it later. For $r = 0$, $q = 1$, we obtain

$$\langle \alpha, \gamma \rangle = 2\frac{|\alpha|}{|\gamma|}cos\theta = 1 \tag{3.3.90}$$

using (3.3.89), where θ is the angle between α and γ. Then, since

$$\langle \gamma, \alpha \rangle = 2\frac{|\gamma|}{|\alpha|}cos\theta \tag{3.3.91}$$

is also satisfied, the relation

$$\frac{\gamma \cdot \gamma}{\alpha \cdot \alpha} = 4cos^2\theta = \langle \gamma, \alpha \rangle \tag{3.3.92}$$

is also concluded. Furthermore, with $\beta = \alpha - \gamma$, we get that the commutation relation between these two is

$$\begin{aligned} &[\chi_\gamma, \chi_\beta] = \chi_\alpha, \quad [\chi_{-\gamma}, \chi_{-\beta}] = -\chi_{-\alpha}, \\ &[\chi_{-\gamma}, \chi_\alpha] = \chi_\beta, \quad [\chi_\gamma, \chi_{-\alpha}] = -\chi_{-\beta}, \\ &[\chi_\alpha, \chi_{-\beta}] = \langle \gamma, \alpha \rangle \chi_\gamma, \quad [\chi_{-\alpha}, \chi_\beta] = -\langle \gamma, \alpha \rangle \chi_{-\gamma}. \end{aligned} \tag{3.3.93}$$

Now we are ready to solve constraint. To solve (3.3.80) and (3.3.81), we first set up ansatz $\Phi_3 = \phi_{3i} H_i (= \text{const})$. In this case, (3.3.80) and (3.3.81) is

$$[A_\mu, \Phi_3] = 0 \tag{3.3.94}$$

$$[A_z, \Phi_3] = 0. \tag{3.3.95}$$

Therefore, $A_{\mu,z}$ also satisfies this constraint if it is a linear combination of Cartan subalgebras. For $A_{\mu,z}$, another solution exists. From (3.3.84), we have

$$[\Phi_3, \chi_\gamma] = \Phi_{3i} \gamma_i \chi_\gamma, \tag{3.3.96}$$

so if we choose Φ_{3i} so that $\Phi_{3i}\gamma_i = 0$, the constraint is satisfied by writing $A_{\mu,z}$ as a linear combination of Cartan subalgebra and $\chi_{\pm\gamma}$. Now, when we define a Cartan subalgebra denoted

$$h_\gamma = \frac{2\gamma_i}{\gamma \cdot \gamma} H_i, \tag{3.3.97}$$

we can write

$$t_1 = \frac{1}{2}i(\chi_\gamma + \chi_{-\gamma}), \quad t_2 = \frac{1}{2}(\chi_\gamma - \chi_{-\gamma}), \quad t_3 = \frac{1}{2}ih_\gamma, \quad y = \frac{1}{2}ih \qquad (3.3.98)$$

for the generator of SU(2)×U(1). Here h is another Cartan subalgebra defined orthogonal to h_γ and normalized to

$$(h, h) = (h_\gamma, h_\gamma) = \frac{4}{\gamma \cdot \gamma}. \qquad (3.3.99)$$

From the above discussion, to satisfy (3.3.80) and (3.3.81), we should choose $\Phi_3 = \phi_{3i} H_i$ and

$$A_{\mu,z} = A^1_{\mu,z}(x, z)t_1 + A^2_{\mu,z}(x, z)t_2 + A^3_{\mu,z}(x, z)t_3 + B_{\mu,z}(x, z)y, \qquad (3.3.100)$$

where $\Phi_{3i}\gamma_i = 0$.

Next, let's solve constraint (3.3.82). Since f_{mnp} is a structure constant of SO(3), constraint is written as

$$[\Phi_3, \Phi_1(x, z)] = -\Phi_2(x, z) \qquad (3.3.101)$$
$$[\Phi_3, \Phi_2(x, z)] = \Phi_1(x, z). \qquad (3.3.102)$$

Here, by defining $\Phi = \Phi_1 + i\Phi_2$, $\tilde{\Phi} = \Phi_1 - i\Phi_2$, the constraint is

$$[\Phi_3, \Phi(x, z)] = i\Phi(x, z) \qquad (3.3.103)$$
$$[\Phi_3, \tilde{\Phi}(x, z)] = -i\tilde{\Phi}(x, z). \qquad (3.3.104)$$

Considering a γ string through α, we can choose $\Phi_{3i}\alpha_i = i$ to keep $\Phi_{3i}\gamma_i = 0$. Then $\Phi_{3i}\beta_i = \Phi_{3i}(\gamma_i - \alpha_i) = -i$, which satisfies constraint (3.3.82) by setting

$$\Phi = \phi_1\chi_\alpha + \phi_2\chi_\beta, \quad \tilde{\Phi} = \tilde{\phi}_1\chi_{-\alpha} + \tilde{\phi}_2\chi_{-\beta}. \qquad (3.3.105)$$

Since $\Phi_{1,2}$ is a real linear combination of generators of Lie algebras, we also find that $\Phi_{1,2}$ is

$$\tilde{\phi}_1 = -\phi_1^*, \quad \tilde{\phi}_2 = -\phi_2^*. \qquad (3.3.106)$$

Finally, Φ_3 is chosen such that $\Phi_{3i}\gamma_i = 0$ and $\Phi_{3i}\alpha_i = i$, so that Φ_3 is determined as follows;

$$\Phi_3 = \frac{1}{2}iZ(h_\alpha + h_\beta), \quad Z = \left(2 - \frac{1}{2}\langle\gamma, \alpha\rangle\right)^{-1}. \qquad (3.3.107)$$

This Φ_3 can be identified with the subgroup U(1) of SU(3) if the basis is well chosen, so the extra gauge field component (ω component) is zero, which satisfies another requirement of the constraint. From the above, the constraint has been solved.

3.4 Heavy and Exotic Baryons in the Sakai-Sugimoto Model

In the Sakai-Sugimoto model, we study the hadron physics by analyzing the $8 + 1$ dimensional flavor gauge theory on the D8 brane in the black 4-brane background. The gauge fields are denoted as A_M^a where $a = 1, ..., 8$ are for the flavor index and $M = 0, ..., 3, z, 6, ..., 9$ for gauge field components. In comparison with actual QCD, they have utilized five components, $A^a_{M=0,...,3,z}$, while the other four were ignored when they derived a five-dimensional gauge theory as discussed in Sect. 2.3.1 [15]. Then the fifth-dimensional degrees of freedom play a role in generating various hadron resonances of light flavors of u, d quarks in the four-dimensional space-time.

In this section, by utilizing a gauge field that lives in the extra dimension, we extend the Sakai-Sugimoto model to the SU(2+1) flavor with heavy quarks. We find that this gauge field transforms to a heavy meson by the method of dimensional reduction proposed by Forgács and Manton (refer to the previous section and also Ref. [3, 34, 35]). Thus, by this method, we obtain an action consisting of a heavy meson and a light meson, where the mass term is supplemented. Starting from this action, we derive the mass formula for the heavy baryon based on the method of collective coordinate quantization of solitons. It is worth noting that our mass formula includes not only conventional states but also exotic states such as P_c states.

3.4.1 *Effective Action with Heavy and Light Mesons*

In this subsection, we introduce heavy flavor to the action of the Sakai-Sugimoto model by using the extra-dimensional gauge field. Therefore, we start the discussion from the 9-dimensional action (2.3.1). The SU(2+1) (light+heavy flavors) gauge fields on the probed D8 brane in the D4 brane background $R^4 \times S^5$ have nine components, $A_M(M = 0\text{–}3, U, \alpha)$, where U is a radial coordinate of S^5 and $\alpha = \psi, \varphi, \theta_1, \theta_2$ angular coordinates of S^5. The gauge fields have also flavor components denoted by the index a, where $A_M = A_M^a \lambda^a/2$ and λ^a are the Gell-Mann matrices. In the previous chapter and Ref. [15, 16], the gauge field components on S^4, A_α, were ignored. In the present work, by regarding A_α^{4-7} among A_α^a as heavy mesons, we try to introduce heavy flavors in the Sakai-Sugimoto (SS) model. Following Forgács-Manton method, this field is transformed into a scalar field corresponding to the heavy meson field by dimensional reduction of the extra dimension S^4 with keeping the gauge field A_α^{4-7}.

Since there are two types of terms in our action, this section is divided into two parts. First, we explain how to reduce the dimensions of the higher dimensional Yang-Mills gauge theory based on the explanation in the previous section. Second, we discuss the Chern-Simons term that we will need.

The Yang-Mills part

In the present analysis, we treat A_ψ and A_φ components as a heavy-meson field to study the system consisting of heavy-mesons and nucleons. A_{θ_1} and A_{θ_2} components are ignored for the minimal use of the extra-dimensional degrees of freedom. To reduce the nine-dimensional theory to four dimensions while preserving the gauge field A_α^{4-7}, we use Forgács-Manton's dimensional reduction method [34].

We start our discussion with the action (2.3.1);

$$S_{D8}^{DBI} \simeq T_8 \left(2\pi^2\alpha'\right)^2 \int d^9x \; e^{-\phi}\sqrt{-\det g} g^{MN} g^{PQ} \mathrm{tr}\left(\frac{1}{4} F_{MP}^{\mathrm{SU(3)}} F_{NQ}^{\mathrm{SU(3)}}\right). \qquad (3.4.1)$$

Here, for the latter discussion, the metric of the D4 black brane is written as the following 9×9 matrix;

$$g^{MN} = \begin{pmatrix} \left(\frac{R}{U}\right)^{3/2} \eta^{\mu\nu} & 0 & 0 \\ 0 & \left(\frac{U}{R}\right)^{3/2} f & 0 \\ 0 & 0 & \left(\frac{U}{R}\right)^{3/2} U^{-2} g^{\alpha\beta}\left(\Omega_4\right) \end{pmatrix}, \qquad (3.4.2)$$

where U is the radial coordinate of S^5, R and U_{KK} characterize the structure of S^5, with $f = 1 - U_{KK}^3/U^3$ and the Minkowski metric, diag $(\eta^{\mu\nu}) = (-1, +1, +1, +1)$. With the generator of SO(3) transfomation ξ_m,

$$\begin{aligned} \xi_1 = \xi_1^M \partial_M &= \cos\varphi \frac{\partial}{\partial\psi} - \cot\psi \, \sin\varphi \frac{\partial}{\partial\varphi}, \\ \xi_2 = \xi_2^M \partial_M &= -\sin\varphi \frac{\partial}{\partial\psi} - \cot\psi \, \cos\varphi \frac{\partial}{\partial\varphi}, \\ \xi_3 = \xi_3^M \partial_M &= \frac{\partial}{\partial\varphi}, \end{aligned} \qquad (3.4.3)$$

which causes rotations around each axis in the (U, ψ, φ) space as Cartesian coordinate, from the discussion of Sect. 3.3, symmetry equation (3.3.4)

$$\left(\partial_M \xi_m^N\right) A_N + \xi_m^N \partial_N A_M = \partial_M W_m + i \left[A_M, W_m\right]. \qquad (3.4.4)$$

finally reduces to the following ansatz

$$W_m = \left(\Phi_3 \frac{\sin\varphi}{\sin\psi}, \ \Phi_3 \frac{\cos\varphi}{\sin\psi}, \ 0 \right), \tag{3.4.5}$$

$$\begin{aligned} A_{\mu,U} &= A_{\mu,U}\left(x^{\nu}, U\right), \\ A_{\psi} &= -\Phi_1\left(x^{\mu}, U\right), \\ A_{\varphi} &= \Phi_2\left(x^{\mu}, U\right)\sin\psi - \Phi_3\cos\psi, \end{aligned} \tag{3.4.6}$$

and constraint

$$\begin{aligned} [\Phi_3, \Phi_1] &= -i\Phi_2, \\ [\Phi_3, \Phi_2] &= i\Phi_1, \\ \left[\Phi_3, A_{\mu,U}\right] &= 0, \end{aligned} \tag{3.4.7}$$

considering the consistency condition (3.3.16), where $\Phi_{1,2}$ are a function of (x^{μ}, U), Φ_3 a constant and we employ a set of ansatz for field configurations.

If we substitute (3.4.5) and (3.4.6) for (3.4.1), we can perform the integration of the higher dimensional manifold S^4 (also refer to Sect. 3.3.3), resulting in a five-dimensional action;

$$\begin{aligned} S_{YM} = \kappa \int d^4x dz \mathrm{tr} & \left[-\frac{1}{2} K^{-1/3} F_{\mu\nu}^2 - K F_{\mu z}^2 \right. \\ & -\frac{4}{9}\left(D_\mu \Phi_m\right)^2 - \frac{4}{9} K^{4/3}\left(D_z \Phi_m\right)^2 \\ & \left. -\frac{16}{81} K^{1/3}\left(i\epsilon_{rst}\Phi_t + [\Phi_r, \Phi_s]\right)^2 \right], \end{aligned} \tag{3.4.8}$$

where $\kappa = N_c\lambda/216\pi^3 = aN_c\lambda$, N_c is a color number, and λ the t'Hooft coupling constant. We use the change of variables between U and z by $U^3/U_{KK}^3 = 1 + z^2 = K$. R and U_{KK} are expressed by Kaluza-Klein mass M_{KK} (2.2.15), which set $M_{KK} = 1$.

Following Sect. 3.3.4, we solve the constraint (3.4.7), then we have the solutions;

$$A_{\mu,z} = A_{\mu,z}^1 \frac{\lambda_1}{2} + A_{\mu,z}^2 \frac{\lambda_2}{2} + A_{\mu,z}^3 \frac{\lambda_3}{2} + A_{\mu,z}^8 \frac{\lambda_8}{2}. \tag{3.4.9}$$

$$\Phi = \frac{1}{2}\begin{pmatrix} 0 & 0 & \phi_1 \\ 0 & 0 & \phi_2 \\ 0 & 0 & 0 \end{pmatrix}, \ \tilde{\Phi} = \frac{1}{2}\begin{pmatrix} 0 & 0 & 0 \\ 0 & 0 & 0 \\ \phi_1^* & \phi_2^* & 0 \end{pmatrix}, \tag{3.4.10}$$

where $\Phi = \Phi_1 + i\Phi_2$, $\tilde{\Phi} = \Phi_1 - i\Phi_2$ and $\phi_{1,2}$ are complex scaler fields [35]. These expressions imply that the gauge fields $A_{\mu,z}$ correspond to light mesons and ϕ_i to heavy mesons. In fact, this situation corresponds to the separation of the brane carrying the heavy flavor from the two light branes in the brane picture. The ϕ

corresponds to the string which connects the light and heavy branes, thus this can be regarded as a heavy meson. It is also discussed in Ref. [39] that the mass of the string on S^4 is heavier. Substituting the light-heavy decomposed fields (3.4.9) and (3.4.10) for the Yang-Mills action (3.4.8), we find

$$\begin{aligned} S_{YM} = \kappa \int d^4x dz \Big\{ \mathrm{tr} \left[-\frac{1}{2} K^{-1/3} F_{\mu\nu}^2 - K F_{\mu z}^2 \right] \\ -\frac{4}{9} \left(D_\mu \phi\right)^\dagger \left(D_\mu \phi\right) - \frac{4}{9} K^{4/3} \left(D_z \phi\right)^\dagger \left(D_z \phi\right) \\ -\frac{16}{81} K^{1/3} \left(\frac{12}{9} - 2\phi^\dagger \phi + \left(\phi^\dagger \phi\right)^2 \right)^2 \Big\}. \end{aligned} \tag{3.4.11}$$

where $\phi^\dagger = \left(\phi_1^*, \phi_2^*\right)$ is a two component isospinor [35]. $F_{\mu\nu,z}$ is the field strength of the SU(2)× U(1) gauge fields (3.4.9), and $D_{\mu,z}$ a covariant derivative.

The Chern-Simon part

We consider the dynamics of (anti) heavy meson fields under the soliton background in the next subsection. It is known that the Wess-Zumino-Witten term plays an important role in the dynamics of the heavy baryon [4, 5, 7], which leads to, for example, the contribution of the attraction (repulsion) that the heavy meson receives from the soliton background. We start with the following Chern-Simons (CS) term,[3] introduced in Ref. [1], because this term leads to the WZW term.

$$\begin{aligned} S_{CS} &= \frac{N_c}{24\pi^2} \int \mathrm{tr} \mathcal{F}^3 \\ \mathrm{tr} \mathcal{F}^3 &= d\omega_5 \left(\mathcal{A}\right) \\ &= d \left[\mathrm{tr} \left(\mathcal{A} \mathcal{F}^2 - \frac{i}{2} \mathcal{A}^3 \mathcal{F} - \frac{1}{10} \mathcal{A}^5 \right) \right], \end{aligned} \tag{3.4.12}$$

where $\mathcal{F}$ is the field strength of $\mathcal{A}$, and the 1-form $\mathcal{A}$ is $\mathcal{A} = \mathcal{A}_M dx^M = A_M dx^M + \hat{A}_M dx^M (M = 0, 1, 2, 3, z, s)$.

The U(1) term of (3.4.12) decomposes

$$\begin{aligned} S_{CS} = \frac{N_c}{24\pi^2} \int \mathrm{tr} F^3 \\ + \frac{N_c}{24\pi^2} \frac{1}{\sqrt{2N_f}} \int \left[3\hat{A} \mathrm{tr} F^2 + \frac{1}{2} \hat{A} \hat{F}^2 \right], \end{aligned} \tag{3.4.13}$$

[3] The Chern-Simon term is defined properly in an odd dimensional space-time. Here following Ref. [1] we call the CS term for $\int \mathrm{tr} \mathcal{F}^3$.

where in the second term we have used the Stokes's theorem to reduce the six-dimensional integral to the five-dimensional one. If we choose $A_z = 0$ gauge, omit massive modes, and integrate over z, the first term is

$$\frac{N_c}{24\pi^2}\int \mathrm{tr}\,F^3 \simeq -\frac{iN_c}{240\pi^2}\int \mathrm{tr}\left(UdU^{-1}\right)^5, \tag{3.4.14}$$

which is nothing but the WZW term [1].

Using the instanton solution of Ref. [40], the Atiyah-Manton approach [41] yields a chiral field U of the following form;

$$U|_{s=0} = \exp\begin{pmatrix} iH(\mathbf{x})\,\hat{\mathbf{x}}\cdot\boldsymbol{\tau}/f_\pi & 0 \\ 0 & 0 \end{pmatrix}, \tag{3.4.15}$$

where $\hat{\mathbf{x}}$ is a unit vector, $\boldsymbol{\tau}$ a Pauli matrix, f_π the decay constant of the pion and the function $H(\mathbf{x})$ is given as

$$\int_{-\infty}^{+\infty} dz' A_z^{cl}\left(\mathbf{x}, z'\right) = H(\mathbf{x})\,\hat{\mathbf{x}}\cdot\boldsymbol{\tau}. \tag{3.4.16}$$

The choice of $s = 0$ corresponds to the boundary of the six-dimensional manifold on which the WZW term is defined. In order for the WZW term to vanish identically in the SU(2) case, we introduce the heavy meson fields $\varphi(\mathbf{x})$ corresponding to the λ_{4-7} components into the chiral fields as follows;

$$U|_{s=0} = \exp\begin{pmatrix} iH(\mathbf{x})\,\hat{\mathbf{x}}\cdot\boldsymbol{\tau}/f_\pi & \varphi(\mathbf{x})/f_H \\ \varphi^\dagger(\mathbf{x})/f_H & 0 \end{pmatrix}, \tag{3.4.17}$$

where f_H is the decay constant of heavy mesons. As we will discuss later, the function $\varphi(\mathbf{x})$ corresponds to the lowest eigenmode of the heavy meson fields ϕ when performed the mode expanded in the fifth z-dimension.

Substituting (3.4.17) for (3.4.14) we find

$$\begin{aligned} &-i\frac{N_c}{240\pi^2}\int \mathrm{tr}\left(U\mathrm{d}U^{-1}\right)^5 \\ &= \frac{iN_c}{f_H^2}\int d^4x\, B^\mu\left(\varphi^\dagger D_\mu\varphi - \left(D_\mu\varphi\right)^\dagger\varphi\right), \end{aligned} \tag{3.4.18}$$

where B_μ is the baryon number current by the soliton,

$$B^\mu = \frac{\epsilon^{\mu\nu\alpha\beta}}{24\pi^2}\mathrm{tr}\left[\left(U_\pi\partial_\nu U_\pi^{-1}\right)\left(U_\pi\partial_\alpha U_\pi^{-1}\right)\left(U_\pi\partial_\beta U_\pi^{-1}\right)\right], \tag{3.4.19}$$

with $U_\pi = \exp\left(iH(\mathbf{x})\,\hat{\mathbf{x}}\cdot\boldsymbol{\tau}/f_\pi\right)$ [5].

The model action

To achieve our aim, we need to introduce a mass term into the action, which is not easy to implement in the Sakai-Sugimoto model. We discussed a way to accomplish this in Sect. 3.2.2, but here, for simplicity, we will supplement the mass term to the action we will use as follows.

$$S = S_{YM} + S_{CS} - m^2 K^{1/3} \phi^\dagger \phi \tag{3.4.20}$$

where the function $K^{1/3}$ is introduced in accordance with (3.4.11) in consideration of the curved nature of the fifth-dimension.

3.4.2 Classical Solutions

In the same way as Sect. 2.4.1, we performe the $1/\lambda$ expansion and obtain a solution for the gauge configuration order by order. In the leading order, the SU(2) $\in$ SU(3) part of the gauge field $A_M^{cl}(\mathbf{x}, z)$ and the U(1) part $\hat{A}_M^{cl}(\mathbf{x}, z)$ are same as the solution (2.4.45). In the next to leading order, the time-components of the SU(2) and U(1) gauge fields are the solution (2.4.45),

The solution of the heavy meson fields ϕ

Now we find a static classical solution of the heavy meson field $\phi(\mathbf{x}, z)$ under the above gauge field background. For this purpose, we first employ mode expansion with a complete set $\{\psi_n(z)\}$ according to Sect. 2.3.2.

$$\phi(\mathbf{x}, z) = \sum_{n=0} \varphi_n(\mathbf{x}) \psi_n(z), \tag{3.4.21}$$

where φ_n are two component isospinors. We can choose an arbitrary complete set $\{\psi_n(z)\}$, and therefore, we choose the one to diagonalize the kinetic and mass terms in the four-dimensional space-time. The eigenvalue equations that such a complete set should satisfy are found according to Sect. 2.3.2 as follows,

$$-\partial_z \left(K^{4/3} \partial_z \psi_n(z)\right) + m^2 K^{1/3} \psi_n(z) = \lambda_n \psi_n(z). \tag{3.4.22}$$

These eigenstates $\psi_n(z)$ correspond to various meson resonances with their eigenvalues regarded as their squared masses. If we consider only the lowest eigenmode, the quadratic terms in ϕ of (3.4.11) become

$$\kappa \int d^4x \left[-\partial_\mu \varphi^\dagger(\mathbf{x}) \partial^\mu \varphi(\mathbf{x}) - m_H^2 \varphi^\dagger(\mathbf{x}) \varphi(\mathbf{x})\right], \tag{3.4.23}$$

where $M = 0, 1, 2, 3, z$, $m_H = \sqrt{\lambda_0}$, and we redefine $\psi = \psi_0$, $\varphi = 2/3\varphi_0$. The mass parameter m is determined such that m_H becomes the heavy meson mass ($D(1870)$, $B(5279)$).

Next, we perform a $1/\lambda$ expansion to derive the equation of motion that $\varphi = 2/3\varphi_0$ must satisfy. We consider the following rescale;

$$\tilde{x}^0 = x^0,\ \tilde{x}^M = \lambda^{1/2} x^M,$$
$$\tilde{\mathcal{A}}_0 = \mathcal{A}_0,\ \tilde{\mathcal{A}}_M = \lambda^{-1/2}\mathcal{A}_M,\ \tilde{\varphi} = \lambda^{-1/2}\varphi, \tag{3.4.24}$$

where $M = 1, 2, 3, z$. In the following calculations, we omit the tilde for simplicity. Then, the action for φ, S_φ becomes to the leading order of $1/\lambda$ expansion

$$S_\varphi \sim aN_c\lambda^1 \int d^4x \left(-\partial_i \varphi^\dagger \partial^i \varphi - \varphi^\dagger \left(\int dz \psi^2 \mathcal{A}_M^2 \right) \varphi \right), \tag{3.4.25}$$

where $\mathcal{A}_M^2$ is proportional to identity matrix by substituting the solution (2.4.45). Therefore, to solve the equation of motion for $\varphi(\mathbf{x})$ we can decompose the two component SU(2) spinor $\varphi(\mathbf{x})$ into $f(\mathbf{x})\chi$, where χ is a two component spinor. Then, the resulting equation of motion for φ is given as

$$\partial_r^2 f + \frac{2}{r}\partial_r f - \left(3 \int dz \frac{\psi^2 (z^2 + r^2)}{(z^2 + r^2 + \rho^2)^2} \right) f = 0. \tag{3.4.26}$$

To solve this equation, we discuss asymptotic behavior, where for simplicity, we rescale the variables as $\xi \to \rho\xi$. First, at $r \to 0$, the third term of (3.4.26) approaches zero, so we know that f has the asymptotic form $f \sim r^{-1}$. Next, we consider the behavior in $r \to \infty$. To do so, multiply (3.4.26) by r^2 to get

$$r^2\partial_r^2 f + r\partial_r f - \left(3 \int dz \psi^2 \frac{(z^2/r^2 + 1)}{(z^2/r^2 + 1 + 1/r^2)^2} \right) f = 0. \tag{3.4.27}$$

If z is small, the integrand of the third term of (3.4.27) becomes

$$\frac{(z^2/r^2 + 1)}{(z^2/r^2 + 1 + 1/r^2)^2} \sim 1.$$

Also if z is large, that term becomes smaller than 1. However, in this case ψ becomes almost zero, so in this region, the third term does not contribute to the equation of motion (3.4.27). Therefore, we can set the third term equals $3f$. From the above, $f|_{r\to\infty}$ satisfies the following equation;

$$r^2\partial_r^2 f + r\partial_r f - 3f = 0.$$

Therefore, the asymptotic behavior at $r \to \infty$ is

$$f \sim r^{\frac{-1-\sqrt{13}}{2}}.$$

We have solved Eq. (3.4.26) numerically satisfying the above asymptotic behaviors. Considering the classical solution of the gauge field and the moduli space parameterized by the collective coordinates of this solution, we will see next that the mass formula of the heavy baryon can be obtained by performing a collective coordinate quantization.

3.4.3 *Quantization*

Collective coordinates

The collective coordinates of the present model consist of those used in Sect. 2.4.2 in addition to that generated by the introduction of the heavy meson.

- Position of the instanton $(\mathbf{X}, Z)$
- Size of the instanton ρ
- SU(2) orientation V
- Two component SU(2) spinor χ

where $(\mathbf{X}, Z)$ and ρ are the position and size of the instanton, respectively, and V the SU(2) matrix corresponding to soliton rotations. The two component SU(2) spinor is introduced as the collective coordinate corresponding to the vibrations of the heavy meson field as follows.

$$\phi = f(\mathbf{x})\,\psi(z)\begin{pmatrix}\chi_1\\ \chi_2\end{pmatrix} = f(\mathbf{x})\,\psi(z)\,\chi. \tag{3.4.28}$$

We give time dependence to the classical solutions from these through the collective coordinates. In the collective coordinate quantization of the gauge theory, we need to be a bit careful [38], which we discussed in Appendix 2.6. Now, for the classical solution we have obtained, we introduce the collective coordinate for the gauge field and for ϕ^{cl} as follows.

$$A_M\left(t, x^N\right) = V A_M^{cl}\left(x^N; X^N(t), \rho(t)\right) V^{-1} - iV\partial_M V^{-1}, \tag{3.4.29}$$
$$\phi\left(t, x^N\right) = V\phi^{cl}\left(x^N; \rho(t), \chi(t)\right), \tag{3.4.30}$$

where $V = V\left(t, x^N\right)$ is an element of the gauge group SU(2). In the $A_0 = 0$ gauge with imposing the Gauss's law:

$$D_M^{cl}\left(\dot{X}^N \frac{\partial}{\partial X^N} A_M^{cl} + \dot{\rho}\frac{\partial}{\partial \rho} A_M^{cl} - D_M^{cl}\Phi\right) = 0, \tag{3.4.31}$$

where $M, N = 1, 2, 3, z$, $\Phi = -iV^{-1}\dot{V}$ and $D_M^{cl} = \partial_M + i\left[A_M^{cl}, \ \right]$. By having the solution of Φ to (3.4.31) [40], spurious motions along the gauge orbits are removed in the collective motions (refer to Appendix 2.6).

Heavy meson field

The action involving the heavy meson fields is written as follows;

$$\begin{aligned} S_\varphi = \ & aN_c \int d^4x \left[\lambda^1 \left(-\partial_i \varphi^\dagger \partial^i \varphi - \varphi^\dagger \left(\int dz \psi^2 \mathcal{A}_M^2\right)\varphi\right)\right. \\ & \left. +\lambda^0 \left(\int dz \psi^2 \left(D_0\varphi\right)^\dagger D_0\varphi - m_H^2 \varphi^\dagger \varphi\right)\right] \\ & +\lambda^0 \frac{iN_c}{f_H^2} \int d^4x B^\mu \left(\varphi^\dagger D_\mu \varphi - \left(D_\mu \varphi\right)^\dagger \varphi\right), \end{aligned} \tag{3.4.32}$$

where the covariant derivative D_0 is defined as $D_0\varphi = \partial_0\varphi + i\mathcal{A}_0\varphi$.

It is convenient to introduce the heavy meson field as [42]

$$\phi = e^{\mp i m_H t}\tilde{\phi} = f(\mathbf{x})\,\psi(z)\,e^{\mp i m_H t}\tilde{\chi}(t), \tag{3.4.33}$$

where $-/+$ correspond to heavy/anti-heavy mesons. Here we perform a $1/m_H$ expansion in addition to the $1/\lambda$ expansion and consider only its leading term. Then, we substitute the solutions (3.4.30) for (3.4.32), the first line of (3.4.32) is zero and the second line becomes

$$\begin{aligned} & \int d^4x dz f^2 \psi^2 \left[\left(D_0\left(V\chi\right)\right)^\dagger D_0\left(V\chi\right) - m_H^2 \chi^\dagger \chi\right] \\ & \simeq 2m_H \int d^4x dz f^2 \psi^2 \tilde{\chi}^\dagger D_0 \tilde{\chi}. \end{aligned} \tag{3.4.34}$$

Finally, the third line becomes

$$\frac{N_c}{\rho^2}\left(\frac{4}{\pi f_H^2}\int dr \sin^2 H \frac{dH}{dr} f^2 - a\int d^3x dz \hat{A}_0^{cl} \psi^2 f^2\right)\tilde{\chi}^\dagger \tilde{\chi}, \tag{3.4.35}$$

because the only the time component of B^{μ} contributes, i.e.

$$B^0 = \frac{1}{2\pi^2}\frac{\sin^2 H}{r^2}\frac{dH}{dr}, \tag{3.4.36}$$

where using the rescale $x^M \to \rho x^M$, $\hat{A}_0^{cl}$ is

$$\hat{A}_0^{cl} = \frac{1}{8\pi^2 a}\frac{1}{\xi^2}\left[1 - \frac{1}{\left(\xi^2+1\right)^2}\right]. \tag{3.4.37}$$

Quantization

By employing the normalization $aN_c \int d^3x dz f^2\psi^2 = 1$, absorbing the coefficient of the kinetic term of $\tilde{\chi}$ and integrating over the space of (x^{μ}, z), in addition to the Lagrangian (2.4.77), we obtain the Lagrangian of the collective coordinate $\tilde{\chi}$

$$\begin{aligned}
&\int dt\left[L_X + L_Z + L_y\right] + \mathcal{O}\left(\lambda^{-1}, m_H^{-1}\right),\\
&L_{\chi} = \pm i\tilde{\chi}^{\dagger}\partial_t\tilde{\chi} \pm A\frac{N_c}{\rho^2}\tilde{\chi}^{\dagger}\tilde{\chi},
\end{aligned} \tag{3.4.38}$$

where we define constant A as follows;

$$\frac{N_c}{\rho^2}\left(\frac{4}{\pi f_H^2}\int dr \sin^2 H\frac{dH}{dr}f^2 - a\int d^3x dz \hat{A}_0^{cl}\psi^2 f^2\right)\tilde{\chi}^{\dagger}\tilde{\chi}, \tag{3.4.39}$$

Before giving the mass formula, we discuss the physical meaning of quantum numbers [4]. In the classical solution, the heavy meson fields have the spin 0 and the isospin 1/2. First, we consider the isospin rotation. By the isospin rotation $g_I = e^{i\boldsymbol{\theta}\cdot\mathbf{I}}$, the gauge fields are transformed into

$$\begin{aligned}
A_M &\to g_I A_M g_I^{-1} - i g_I \partial_M g_I^{-1}\\
&= \left(g_I V\right) A_M^{cl}\left(g_I V\right)^{-1} - i\left(g_I V\right)\partial_M\left(g_I V\right)^{-1}.
\end{aligned} \tag{3.4.40}$$

On the other hand, the heavy meson field transforms

$$V\tilde{\chi} \to g_I V\tilde{\chi}. \tag{3.4.41}$$

Therefore, V carries the isospin, and V and $\tilde{\chi}$ have the following transformation properties:

$$\begin{cases} V \to g_I V \\ \tilde{\chi} \to \tilde{\chi}. \end{cases} \tag{3.4.42}$$

Second, we consider the spatial rotation. When the gauge transformation which is equivalent to the spatial rotation is written as $g_J = e^{i\theta\cdot\mathbf{J}}$, spatial rotation act the gauge field as follows:

$$\begin{aligned} &A_M\left(t, R_{NP}x^P\right) \\ &= VA_M^{cl}\left(R_{NP}x^P; R_{NP}X^P\right)V^{-1} - iVV^{-1} \\ &= \left(Ve^{-i\theta\cdot\mathbf{I}}\right)A_M^{cl}\left(x^N; X^N\right)\left(Ve^{i\theta\cdot\mathbf{I}}\right)^{-1} - i\left(Ve^{-i\theta\cdot\mathbf{I}}\right)\partial_M\left(Ve^{i\theta\cdot\mathbf{I}}\right)^{-1}, \end{aligned} \tag{3.4.43}$$

where the hedgehog like structure relates the spatial rotation to isospin rotation, and so the spatial rotation is expressed by g_I. Also, the scalar field is transformed into

$$V\tilde{\phi}(t,x) \to V\tilde{\phi}\left(t, R_{MN}x^N\right) = Ve^{i\theta\cdot\mathbf{J}}\tilde{\phi} = Ve^{-i\theta\cdot\mathbf{I}} f\psi e^{i\theta\cdot\mathbf{T}}\tilde{\chi}(t), \tag{3.4.44}$$

where $\mathbf{T} = \mathbf{J} + \mathbf{I}$ is the grand spin operator. Therefore, V and χ have the following transformation properties:

$$\begin{cases} V \to Ve^{-i\theta\cdot\mathbf{I}} \\ \tilde{\chi} \to e^{i\theta\cdot\mathbf{T}}\tilde{\chi}. \end{cases} \tag{3.4.45}$$

From the above, after doing the collective rotation, $\tilde{\chi}$ has the spin 1/2 and the isospin 0. Thus, we should quantize $\tilde{\chi}$ as fermions:

$$\left\{\tilde{\chi}_i, \tilde{\chi}_j^\dagger\right\} = \tilde{\chi}_i\tilde{\chi}_j^\dagger + \tilde{\chi}_j^\dagger\tilde{\chi}_i = \delta_{ij}. \tag{3.4.46}$$

3.4.4 Mass Formula

By correcting the term proportional to $1/\rho^2$ in (2.4.77), our Hamiltonian has the same form as in Ref. [40]. Therefore, we perform the same quantization procedure, which leading to the following mass formula,

$$\begin{aligned} M = \; & M_0 + \left(N_Q + N_{\overline{Q}}\right)m_H \\ & + \sqrt{\frac{(l+1)^2}{6} + \frac{2N_c^2}{15}\left(1 - \frac{40a\pi^2 A}{N_c}\left(N_Q - N_{\overline{Q}}\right)\right)}\,M_{KK} \\ & + \frac{2\left(n_\rho + n_Z\right) + 2}{\sqrt{6}}M_{KK}, \end{aligned} \tag{3.4.47}$$

Table 3.3 Parameters in our model

M_0(MeV)	M_{KK}(MeV)	m/M_{KK}	m/M_{KK}	f_π/M_{KK}
–572	500	4.385 (Charm)	10.62 (Bottom)	0.122

where M_0 is the instanton mass, $N_{Q/\overline{Q}}$ the number of heavy/anti-heavy mesons. The numerical constant, A, is the new term added to the mass formula of Ref. [40] in our analysis.

The spin of a heavy baryon is represented by the sum of the spin of the instantons and havey mesons. The instanton spin and isospin are both $l/2$ from the hedgehog structure. Also, as mentioned in Sect. 2.4.4, when n_Z is even or odd, the wave function of n_Z has even and odd parities. Note that $\tilde{\phi} = f\psi\tilde{\chi}$ has parity even.

Parameters in or mass are (M_0, M_{KK}, m, f_π) and are shown in Table 3.3. Also, m is determined so that m_H is the mass $(D(1870), B(5279))$ of the heavy meson. The pion decay constant is $f_\pi = 61.2$ MeV which is about 30% smaller than the experimental value 93.2 MeV [43, 44]. The heavy meson decay constant f_H was determined to be $f_D/f_\pi = 1.7$ and $f_B/f_\pi = 1.6$ in the charm and bottom baryon analyses, respectively, with reference to Ref. [43]. For the Kalza-Klein mass, we used $M_{KK} = 500$ MeV, which reproduces the nucleon and $\Delta(1232)$ mass splitting, following Ref. [40]. Having these inputs, there is only one free parameter M_0 which is fixed to reproduced the mass of $\Lambda_c(2286)$. We see that our mass formula differs from Ref. [12] by a term proportional to A. This term depends on the heavy meson decay constant f_H, while there is no such parameter dependence in the mass formula of Ref. [12]. From (3.4.39) with the decay constant values as in Table 3.3, we find $A = 0.078$ for charm and $A = 3.7$ for bottom sectors, respectively.

Results are summarized in Table 3.4.

These results have some characteristic properties as follows.

- As in Ref. [40] and Table 2.3, this mass formula shows degeneracy of the Roper resonance and negative parity resonance, which is consistent with the experimental data in the light flavor sector, while this feature is difficult to explain in the quark model. This feature is also generalized to the hyperon case [45], hence it is an interesting question whether this feature exists when extended to charms and bottoms. Possible candidates are $\Lambda_c(2765)$ and $\Lambda_b(6072)$, while their spin and parity are to be determined. In particular, since it has been argued that the excitation energy of the Roper resonance is flavor independent, our results are consistent [46] with this context.
- If we expand the mass formula (3.4.47) by $1/N_c$, the mass splitting of Λ_c and Σ_c is proportional to $1/N_c$. This splitting is related to the spin-spin interaction, and the N_c dependency is consistent to that of the $1/N_c$ expansion scheme.
- Because we have included only the leading terms of $1/m_H$ expansion, we have obtained the heavy quark symmetry (HQS) singlet $\Lambda_{c,b}(0\frac{1}{2}^+)$ and the degenerate doublet $\Sigma_{c,b}(1\frac{1}{2}^+, 1\frac{3}{2}^+)$. On the other hand, the lowest $\Lambda_{c,b}(0\frac{1}{2}^-)$ and $\Lambda_{c,b}(0\frac{3}{2}^-)$,

Table 3.4 Predictions of our mass formula for the charmed and bottomed baryons in comparison with experimental data where available. Masses of heavy quark doublet, for instance Σ_c and Σ_c^*, are degenerate in the heavy meson limit $m_H \to \infty$

B	IJ^P	l	n_ρ	n_z	N_Q	$N_{\overline{Q}}$	Our model/MeV	Exp./MeV
Λ_c	$0\frac{1}{2}^+$	0	0	0	1	0	[2286]	2286
Σ_c	$1\frac{1}{2}^+$	2	0	0	1	0	2523	2453
	$1\frac{3}{2}^+$	2	0	0	1	0	2523	2520
Λ_c^*	$0\frac{1}{2}^-$	0	0	1	1	0	2694	(2595)
	$0\frac{1}{2}^+$	0	1	0	1	0	2694	(2765)
Σ_c^*	$1\frac{1}{2}^-, 1\frac{3}{2}^-$	2	0	1	1	0	2931	–
	$1\frac{1}{2}^+, 1\frac{3}{2}^+$	2	1	0	1	0	2931	–
P_c	$\frac{1}{2}\frac{1}{2}^-, \frac{1}{2}\frac{3}{2}^-$	1	0	0	1	1	4255	4312/4380/4440, 4457
P_c^*	$\frac{1}{2}\frac{1}{2}^-, \frac{1}{2}\frac{3}{2}^-$	1	0	1	1	1	4664	–
	$\frac{1}{2}\frac{1}{2}^+, \frac{1}{2}\frac{3}{2}^+$	1	1	0	1	1	4664	–
Λ_b	$0\frac{1}{2}^+$	0	0	0	1	0	5676	5620
Σ_b	$1\frac{1}{2}^+$	2	0	0	1	0	5919	5810
	$1\frac{3}{2}^+$	2	0	0	1	0	5919	5830
Λ_b^*	$0\frac{1}{2}^-$	0	0	1	1	0	6084	5912
	$0\frac{1}{2}^+$	0	1	0	1	0	6084	(6072)
Σ_b^*	$1\frac{1}{2}^-, 1\frac{3}{2}^-$	2	0	1	1	0	6327	–
	$1\frac{1}{2}^+, 1\frac{3}{2}^+$	2	1	0	1	0	6327	–
P_b	$\frac{1}{2}\frac{1}{2}^-, \frac{1}{2}\frac{3}{2}^-$	1	0	0	1	1	11070	–
P_b^*	$\frac{1}{2}\frac{1}{2}^-, \frac{1}{2}\frac{3}{2}^-$	1	0	1	1	1	11480	–
	$\frac{1}{2}\frac{1}{2}^+, \frac{1}{2}\frac{3}{2}^+$	1	1	0	1	1	11480	–

which correspond to λ mode in a quark model, do not exist in the present model. This is because the present analysis considers only instanton excitations, which in terms of the quark model is related to the ρ mode. The negative parity lowest mode $\Lambda_{c,b}(0\frac{1}{2}^-)$ and $\Lambda_{c,b}(0\frac{3}{2}^-)$ corresponding to λ mode should be described by considering the bound state of the instanton as a nucleon and the second lowest eigenmode of the mode expansion (3.4.22). These are future work.

- Empirically, the mass splitting of Λ_c and Λ_c^* is about twice larger than that of Λ_c and Σ_c. In the present study, the value of A plays an important role to make this order of baryon masses. In particular, for $\Delta_{\Sigma_c-\Lambda_c} \equiv M(\Sigma_c) - M(\Lambda_c)$ and $\Delta_{\Lambda_c^*-\Lambda_c} \equiv M(\Lambda_c^*) - M(\Lambda_c)$, we have $\Delta_{\Sigma_c-\Lambda_c} < \Delta_{\Lambda_c^*-\Lambda_c}$ in accordance with the experimental data, while the formula in Ref. [12] results in the reversed relation. Let B be

$$B = 1 - \frac{40a\pi^2 A}{N_c}. \tag{3.4.48}$$

For $B = 0$, we find $\Delta_{\Sigma_c-\Lambda_c} = \Delta_{\Lambda_c^*-\Lambda_c}$. As A becomes smaller (i.e. B becomes larger), $\Delta_{\Sigma_c-\Lambda_c}$ becomes larger, and at some point, $\Delta_{\Lambda_c^*-\Lambda_c}$ equals $2\Delta_{\Sigma_c-\Lambda_c}$. Recently, in Ref. [13], they have realized the same mass ordering as ours by the correction to the mass formula of Ref. [12], as a result of their analysis of the subleading term up to $1/m_H$.

- Our mass formula can also predict the recently reported $P_c(4312/4380/4440, 4457)$ states. Similarly, we predict the masses of the hidden bottomed pentaquark states, which have not yet been observed (denoted as P_b in this paper). However, due to the instanton hedgehog structure, our mass formula cannot generate $\frac{1}{2}\frac{5}{2}^+$.

3.4.5 Conclusion

In this section we discuss how to analyze the heavy baryon by the Sakai-Sugimoto model and derive the mass formula. We describe the heavy baryon as a bound state of instantons and heavy mesons corresponding to nucleons by interpreting the extra-dimensional component of the gauge field as a heavy meson, which was ignored in Ref. [15]. The heavy baryon is described as the bound state of the instanton and heavy meson corresponding to the nucleon. In this process, we obtained the action consisting of the light meson and heavy meson fields that form the instanton by using the Forgács-Manton method [34, 35] as a method of dimensional reduction while keeping the extra dimensional component of the gauge field corresponding to the heavy meson. To obtain the mass formula, we introduced a new collective coordinate for the vibrations of the heavy meson and performed the quantization of the system according to the method of collective coordinate quantization. When quantizing our model, as in Ref. [4], heavy mesons behave as heavy quarks, which was referred to as transmutation of quantum numbers in the intrinsic frame of the hedgehog instanton [12]. Finally, we compared the obtained mass spectra with the experimental data and confirmed that our results are consistent with the experiments.

In our model, we have considered the limit of the large N_c and the t'Hooft coupling λ as in [40], and also took the limit of large m_H. We have treated the only leading terms of $1/m_H$, so we have obtained the HQS singlet $\Lambda_{c,b}(0\frac{1}{2}^+)$ and the doublet $\Sigma_{c,b}(1\frac{1}{2}^+, 1\frac{3}{2}^+)$. Also, our mass formula has yielded the degenerate Roper like and odd parity excitations. Moreover, we have realized the mass ordering $\Delta_{\Sigma_c-\Lambda_c} < \Delta_{\Lambda_c^*-\Lambda_c}$, which is failed to reproduce in Ref. [10] in accordance with the experimental data. Furthermore, our model has hidden charmed pentaquark states $P_c(4312/4380/4440, 4457)$ reported recently [47, 48]. Similarly, we have predicted the masses of hidden bottomed pentaquark states not yet observed.

As a further development, the analysis performed for nucleon resonance in Sect. 4.3 may be applied to heavy baryons. The Forgács-Manton method [34] used in this study may also be applied to the study of neutron stars with hyperons. Our recent work has shown that this method can be useful in considering ansatz for introducing s quarks in the study of neutron stars by holographic QCD. This is another interesting future work.

3.5 Appendix: Lie Derivative

In this section, we explain the Lie derivative according to the textbook [49]. Now, the vector field at each point on the manifold M is written as $X \in \mathscr{X}(M)$ and the 1-form is written as $\omega \in \Omega^1(M)$. A curved line is defined on the manifold M such that this vector field is a tangent vector. With x as a certain point, the infinitesimal transformation along the curve generated by X can be expressed as $\sigma_\epsilon^\mu(x) = x^\mu + \epsilon X^\mu(x)$ with the infinitesimal quantity ϵ (When there is no subscript of μ, the argument is valid regardless of whether or not coordinates are introduced.). The derivative of 1-form ω along this change is called the Lie derivative. However, it does not make sense to simply compare the 1-forms in x and σx, because they are 1-forms belonging to different cotangent spaces.

Now, we define the map $\sigma_\epsilon : x \to \sigma_\epsilon(x)$, then this map induces a map $(\sigma_\epsilon)_* : T_x M \to T_{\sigma_\epsilon(x)} M$ in the tangent space. As a result, the map $(\sigma_\epsilon)^* : T^*_{\sigma_\epsilon(x)} M \to T^*_x M$ is induced in the cotangent space. This mapping $(\sigma_\epsilon)^*$ is called the pullback of σ_ϵ. From this, the 1-form on $T^*_{\sigma_\epsilon(x)} M$ in $\sigma_\epsilon(x)$ (written $\omega|_{\sigma_\epsilon(x)}$ in the following) can be pulled back (written $(\sigma_\epsilon)^*\omega|_{\sigma_\epsilon(x)}$) on $T^*_x M$. Thus, a meaningful derivative is defined by comparing $(\sigma_\epsilon)^*\omega|_{\sigma_\epsilon(x)}$ and $\omega|_x$. Concretely, we can define it as

$$\mathbf{L}_X \omega = \lim_{\epsilon \to 0} \frac{1}{\epsilon} \left[(\sigma_\epsilon)^* \omega|_{\sigma_\epsilon(x)} - \omega|_x \right], \tag{3.5.1}$$

where the operator of the Lie derivative along X is written as $\mathbf{L}_X$. Let's transform the equation on the right hand side into a convinient form by introducing coordinates. Let us write $X = X^\mu \partial/\partial x^\mu$ for the vector field on $T_x M$ and $\omega = \omega_\mu dx^\mu$ for the 1-form on $T^*_x M$. Then $\sigma_\epsilon(x)$ has the coordinate $x^\mu + \epsilon X^\mu(x)$, and the 1-form on $T^*_{\sigma_\epsilon(x)} M$ is transformed as follows;

$$\begin{aligned} \omega|_{\sigma_\epsilon(x)} &= \omega_\mu(x^\nu + \epsilon X^\nu) d(x^\mu + \epsilon X^\mu) \\ &= \big(\omega_\mu(x) + \epsilon X^\nu(x) \partial_\nu \omega_\mu(x) + \mathcal{O}(\epsilon^2)\big) d(x^\mu + \epsilon X^\mu). \end{aligned} \tag{3.5.2}$$

If this is pulled back over $T^*_x M$ by $(\sigma_\epsilon)^*$, it is computed to be

$$
\begin{aligned}
(\sigma_\epsilon)^*\omega|_{\sigma_\epsilon(x)} &= \big(\omega_\mu(x) + \epsilon X^\nu(x)\partial_\nu\omega_\mu(x) + \mathcal{O}(\epsilon^2)\big)\frac{d(x^\mu + \epsilon X^\mu)}{dx^\rho}dx^\rho \\
&= \big(\omega_\mu(x) + \epsilon X^\nu(x)\partial_\nu\omega_\mu(x) + \mathcal{O}(\epsilon^2)\big)\big(\delta^\mu_\rho + \epsilon\partial_\rho X^\mu\big)dx^\rho \\
&= \omega_\mu dx^\mu + \epsilon\big(X^\nu(x)\partial_\nu\omega_\mu(x) + \partial_\mu X^\nu(x)\omega_\nu(x)\big)dx^\mu + \mathcal{O}(\epsilon^2). \quad (3.5.3)
\end{aligned}
$$

From the above, the Lie derivative of $\omega_\mu dx^\mu$ by $X^\nu\partial/\partial x^\nu$ is expressed as

$$
\mathbf{L}_X\omega = \big(X^\nu(x)\partial_\nu\omega_\mu(x) + \partial_\mu X^\nu(x)\omega_\nu(x)\big)dx^\mu. \quad (3.5.4)
$$

In this doctoral thesis, when we say that a field is symmetric or invariant with respect to the transformations generated by the vector field X, we mean that this Lie derivative is zero (for gauge fields, this restriction is more relaxed, as in (3.3.3)). We can correspond X to ξ and ω to the gauge field $A = A_\mu dx^{mu}$. From the above definition, it is understood that symmetric fields are, for example, those shown in Figs. 3.3 and 3.4.

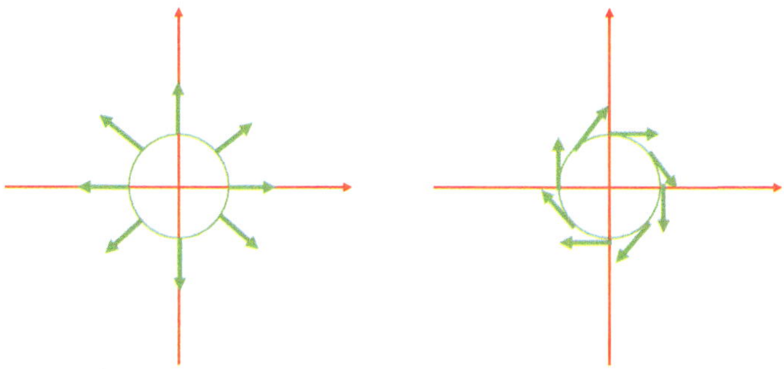

Fig. 3.3 Rotational symmetric fields

Fig. 3.4 Translation-symmetric fields

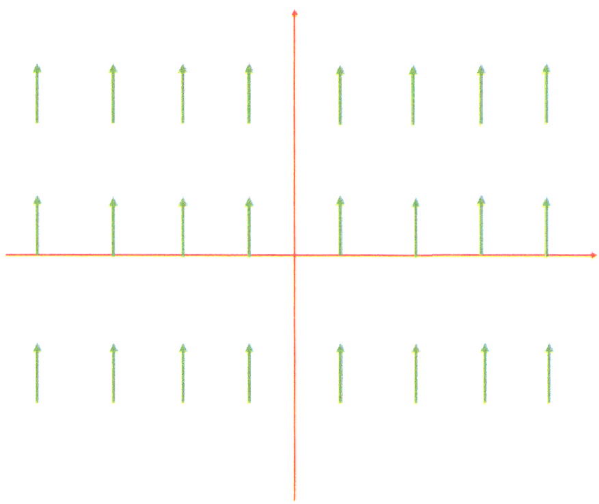

3.6 Appendix: Chern-Simons Term

3.6.1 CS Term

In this chapter, we obtained the mass formula for baryons in the flavor SU(3) Sakai-Sugimoto model, yet we proceeded with the analysis assuming that a constraint term (3.2.47), which does not actually derive from the CS term used in the Ref. [15, 16]. So far, there have been studies to improve the CS term so that the SakaiSugimoto model produces a constraint term [1, 2]. In the following, we will refer to [1, 2] and explain the issues regarding the CS term, including a discussion of the constraint term in the Sakai-Sugimoto model.

First, let us summarize a few things that need to be clarified when considering the problems associated with the CS term in the Sakai-Sugimoto model. Consider a five-dimensional YM-CS theory in the Sakai-Sugimoto model defined on a manifold M_5. The boundary of this manifold M_5 in the Sakai-Sugimoto model is the four-dimensional manifold $M_4^{(\pm\infty)}$ at $z \to \pm\infty$. That is,

$$\partial M_5 = M_4^{(+\infty)} \cup \left(- M_4^{(-\infty)} \right). \tag{3.6.1}$$

The minus sign means that the orientations are reversed. Here, the field defined on $M_4^{(\pm\infty)}$ is written as $\hat{A}_\pm = A|_{z\to\pm\infty}$. These are the gauged external fields corresponding to the chiral symmetry $U(N_f)_L \times U(N_f)_R$ of QCD, respectively. The CS term is invariant under gauge transformations that act trivially on this boundary, but it changes under gauge transformations that act nontrivially on the boundary, which leads to the chiral anomaly in QCD.

In fact, the variation caused by the infinitesimal gauge transformation $\delta_\Lambda A = d\Lambda + [A, \Lambda] = D_\Lambda A$ of CS term

$$S_{\mathrm{D8}}^{\mathrm{CS}} = \frac{iN_c}{24\pi^2} \int_{M_4\times\mathbb{R}} \omega_5(A) \tag{3.6.2}$$

$$\omega_5(A) = tr\left(AF^2 - \frac{1}{2}A^3F + \frac{1}{10}A^5 \right) \tag{3.6.3}$$

$$= tr\left(AdAdA + \frac{3}{2}A^3dA + \frac{3}{5}A^5 \right) \tag{3.6.4}$$

used in Ref. [15, 16] (In this section, we use the anti-Hermitian $U(N_f)$ gauge field $A^\dagger = -A$, so i appears in the coefficients of the CS term and the CS 5-form notation is different.) become

$$\delta_\Lambda S_{CS} = C \int_{M_4\times\mathbb{R}} \delta_\Lambda \omega_5(A)$$
$$= C\Big(\int_{M_4^{+\infty}} \omega_4^1(\hat{\Lambda}_+, \hat{A}_+) - \int_{M_4^{-\infty}} \omega_4^1(\hat{\Lambda}_-, \hat{A}_-)\Big), \tag{3.6.5}$$

which is the chiral anomaly of QCD (see also the textbook on [50]). Here, with

$$C = \frac{iN_c}{24\pi^2} \tag{3.6.6}$$
$$\omega_4^1(\Lambda, A) = \mathrm{tr}\Big(\Lambda d\Big(AdA + \frac{1}{2}A^3\Big)\Big) \tag{3.6.7}$$
$$\hat{\Lambda}_\pm = \Lambda|_{z\to\pm\infty}, \tag{3.6.8}$$

we wrote in differential form and omitted the wedge product $\wedge$ and further used the relation,

$$\delta_\Lambda \omega_5(A) = d\omega_4^1(\Lambda, A) + \mathcal{O}(\Lambda^2) \tag{3.6.9}$$

and Stokes' theorem.

We now define the gauge field on the boundary $M_4^{(\pm\infty)}$ as $\hat{A}_\pm = A|_{z\to\pm\infty}$. However, as long as we use such a globally well-defined gauge field, we see that the baryon number is necessarily zero.

Different from the coordinate z, $|x^\mu| \to \infty$ is not considered a boundary, so we compactify it to $M_4 = S^1 \times S^3$ to avoid confusion, where S^1 is time and S^3 space. The baryon number is the instanton number given as

$$N_B = \frac{1}{8\pi^2} \int_{S^3\times\mathbb{R}} \mathrm{tr}(F^2), \tag{3.6.10}$$

where the space z extends is denoted as $\mathbb{R}$. Here, from

$$\mathrm{tr}(F^2) = d\omega_3(A) \tag{3.6.11}$$
$$\omega_3(A) = \mathrm{tr}\Big(AF - \frac{1}{3}A^3\Big) = \mathrm{tr}\Big(AdA + \frac{2}{3}A^3\Big), \tag{3.6.12}$$

it can be transformed to

$$N_B = \frac{1}{8\pi^2}\Big(\int_{S^3} \omega_3(A)|_{z\to+\infty} + \int_{S^3} \omega_3(A)|_{z\to-\infty}\Big). \tag{3.6.13}$$

From this formula, we conclude that if $A|_{z\to\pm\infty} = 0$, then the baryon number N_B is zero. Therefore, if we define $\hat{A}_\pm = A|_{z\to\pm\infty}$, we can always set $\hat{A}_\pm = 0$ by a gauge transformation, so we can only describe the case $N_B = 0$, as long as we consider a globally well-defined gauge field. This result does not suit our purpose of trying to analyze baryons in instantons. However, in gauge theory, only the field strength needs to be globally well-defined, not the gauge field. Keeping this in mind, we can construct a solution for $N_B > 1$, which also allows us to use the BPST Instanton solution.

To obtain a gauge configuration with $N_B > 1$, we divide the space-time manifold M_5 into two or more patches

$$
\begin{aligned}
&M_5 = M_5^- \cup M_5^+ \\
&M_5^{\pm} = \{(x, z) \in M_5 | \pm z > -\epsilon\} \\
&M_5^- \cap M_5^+ \simeq M_4^{(0)} \times (-\epsilon, +\epsilon) \\
&M_4^{(0)} = \{(x^\mu, z) \in M_5 | z = 0\} \simeq S^1 \times S^3
\end{aligned}
\tag{3.6.14}
$$

and define a gauge field on each of these patches (Fig. 3.5). For the field strength to be globally well defined, the gauge field should be connected to

$$
A_+ = A_-^h = hA_-h^{-1} + hdh^{-1} \tag{3.6.15}
$$

by $h \in U(N_f)$ at $M_5^- \cap M_5^+ \simeq M_4^{(0)} \times (-\epsilon, +\epsilon)$ where these two patches overlap. Also, the gauge transformation should be defined by

$$
A_\pm \to A_\pm^{g_\pm} = g_\pm A_\pm g_\pm^{-1} + g_\pm d g_\pm^{-1}, \quad h \to g_+ h g_-^{-1} \tag{3.6.16}
$$

with $g_\pm \in U(N_f)$.

Here, different from the previous case, we define the gauge field at the boundary of $z \to \pm\infty$ by

$$
\hat{A}_\pm = A_\pm|_{z\to\pm\infty}. \tag{3.6.17}
$$

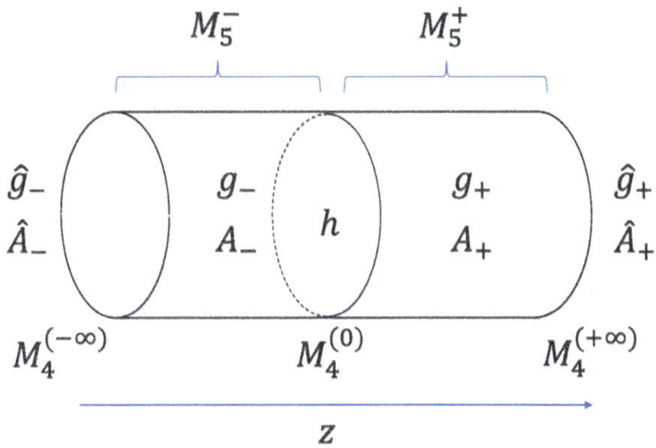

Fig. 3.5 The space-time manifold in the Sakai-Sugimoto model

Furthermore, the gauge transformation $g_\pm \in U(N_f)$ acts like

$$\hat{g}_\pm = g_\pm|_{z\to\pm\infty} \tag{3.6.18}$$

to the boundary $z \to \pm\infty$, which corresponds to the gauged elements of the chiral symmetry, $(\hat{g}_-, \hat{g}_+) \in U(N_f)_L \times U(N_f)_R$.

This allows for a gauge configuration of $N_B > 1$. Denoting the space manifold in $Z = -\infty, 0, +\infty$ as $S^3_{(-\infty)}, S^3_{(0)}, S^3_{(+\infty)}$, the baryon number is transformed into the following form

$$\begin{aligned} N_B =& \frac{1}{8\pi^2}\Big(\int\limits_{S^3\times(-\infty,0]} d\omega_3(A_-) + \int\limits_{S^3\times[0,+\infty)} d\omega_3(A_+)\Big) \\ =& \frac{1}{8\pi^2}\Big(\int\limits_{S^3_{(+\infty)}} \omega_3(A_+) + \int\limits_{S^3_{(-\infty)}} \omega_3(A_-) \\ &+ \int\limits_{S^3_{(0)}} \omega_3(A_+) + \int\limits_{-S^3_{(0)}} \omega_3(A_-)\Big), \end{aligned} \tag{3.6.19}$$

with the first line unchanged and the second line added. Here, if we use

$$\omega_3(A^g) = \omega_3(A) - \frac{1}{3}\mathrm{tr}((gdg^{-1})^3) - d\mathrm{tr}(dg^{-1}gA) \tag{3.6.20}$$

with $\hat{A}_\pm = 0$, we obtain the formula

$$N_B = \frac{1}{24\pi^2}\int\limits_{S^3} \mathrm{tr}((hdh^{-1})^3)|_{z=0}, \tag{3.6.21}$$

which is nothing but the definition of the mapping degree of $h : S^3 \to SU(N_f)$, namely $\pi_3(U(N_f)) = \mathbb{Z}$.

Vanishing of the constraint term

In the following, we show that in the CS term used in the Sakai-Sugimoto model, the constraint term that plays an important role in obtaining the baryon spectrum does not appear, as explained in Sect. 3.1. From now on, we will use Hermitian gauge fields. First we give some formulas related to the CS 5-form. Under the gauge transformation $\mathcal{A} \to \mathcal{A}^V = V(\mathcal{A} - id)V^{-1}$ of the gauge field $\mathcal{A}$ (decomposed as (2.4.14)) of the $U(N_f)$ gauge group, we deform it as

$$\omega_5(\mathcal{A}^V) = \omega_5(\mathcal{A}) + \frac{1}{10}\mathrm{tr}L^5 + d\alpha_4(L.\mathcal{A}), \tag{3.6.22}$$

where it is defined to be

$$L = -iV^{-1}dV \tag{3.6.23}$$

$$\alpha_4(L, \mathcal{A}) = \frac{1}{2}\mathrm{tr}\Big[L(\mathcal{A}\mathcal{F} + \mathcal{F}\mathcal{A} - i\mathcal{A}^3) + \frac{i}{2}L\mathcal{A}L\mathcal{A} - iL^3\mathcal{A}\Big]. \tag{3.6.24}$$

Also, in an arbitrary infinitesimal transformation $\mathcal{A} \to \mathcal{A} + \delta\mathcal{A}$ of this gauge field, we deform it as

$$\omega_5(\mathcal{A} + \delta\mathcal{A}) = \omega_5(\mathcal{A}) + 3\mathrm{tr}(\delta\mathcal{A}\mathcal{F}^2) + d\beta(\delta\mathcal{A}, \mathcal{A}) + \mathcal{O}((\delta\mathcal{A})^2), \tag{3.6.25}$$

where $\beta(\delta\mathcal{A}, \mathcal{A})$ is defined to be

$$\beta(\delta\mathcal{A}, \mathcal{A}) = \mathrm{tr}\Big[\delta\mathcal{A}\Big(\mathcal{F}\mathcal{A} + \mathcal{A}\mathcal{F} - \frac{i}{2}\mathcal{A}^3\Big)\Big]. \tag{3.6.26}$$

Using the above formula, the CS 5-form in the gauge field with time dependence by the collective coordinate is given in the form

$$\begin{aligned}\omega_5(\mathcal{A}) &= \omega_5\big((\mathcal{A}^{cl} + \Phi dt)^W\big)\\ &= \omega_5(\mathcal{A}^{cl} + \Phi dt) + \frac{1}{10}\mathrm{tr}(-iW^{-1}\dot{W}dt)^5 + d\alpha_4(-iW^{-1}\dot{W}dt, \mathcal{A}^{cl} + \Phi dt)\\ &= \omega_5(\mathcal{A}^{cl}) + 3\mathrm{tr}\big(\Phi dt(\mathcal{F}^{cl})^2\big) + d\beta(\Phi dt, \mathcal{A}^{cl}) + d\alpha_4(-iW^{-1}\dot{W}dt, \mathcal{A}^{cl})\\ &= \omega_5(\mathcal{A}^{cl}) + 3\mathrm{tr}\big(\Phi dt(F^{cl})^2\big) + d\beta(\Phi dt, A^{cl}) + d\alpha_4(-iW^{-1}\dot{W}dt, A^{cl}).\end{aligned} \tag{3.6.27}$$

In the last line of the deformation, we used $\hat{A}^{cl}_M(x; X^\alpha(t)) = \hat{F}^{cl}_{MN}(x; X^\alpha(t)) = 0$ from the solution we already obtained (3.2.20). Therefore, the CS 5-form is transformed by giving time dependence to the collective coordinates, leading to three terms other in addition to $\omega_5(\mathcal{A}^{cl})$. However, the contributions of these three terms cancel each other out, eventually resulting in $\omega_5(\mathcal{A}) = \omega_5(\mathcal{A}^{cl})$. Let us explain this in the following.

We first consider the term $3\mathrm{tr}\big(\Phi dt(F^{cl})^2\big)$. Since A^{cl}_M is an embedded SU(2) BPST instanton solution (3.2.8), we get

$$(F^{cl})^2 = \frac{1}{2}\mathcal{P}_2\mathrm{tr}(F^{cl})^2. \tag{3.6.28}$$

Since only t_8 of the SU(3) generators survive when multiplied by $\mathcal{P}_2$ and take a trace, we obtain the relation

$$\begin{aligned}\operatorname{tr}\bigl(\Phi(F^{cl})^2\bigr) &= \operatorname{tr}\bigl(\chi^a\Phi_a\mathcal{P}_2\bigr)\frac{1}{2}\operatorname{tr}\bigl(F^{cl}\bigr)^2 \\ &= \frac{1}{2\sqrt{3}}\operatorname{tr}\bigl(F^{cl}\bigr)^2,\end{aligned} \tag{3.6.29}$$

with

$$\operatorname{tr}(\Phi\mathcal{P}_2) = \frac{1}{\sqrt{3}}\chi^8(t) \tag{3.6.30}$$

From

$$\mathcal{P}_2 = \begin{pmatrix}1&0&0\\0&1&0\\0&0&0\end{pmatrix} = \frac{2}{\sqrt{3}}t_8 + \frac{2}{3}\mathbf{1}_3, \quad t_8 = \frac{1}{2\sqrt{3}}\begin{pmatrix}1&0&0\\0&1&0\\0&0&-2\end{pmatrix} \tag{3.6.31}$$

it is computed that

$$\begin{aligned}\frac{N_c}{24\pi^2}\int\limits_{M_5=\mathbb{R}\times M_4} 3\operatorname{tr}(\Phi dt(F^{cl})^2) &= \frac{N_c}{24\pi^2}\frac{\sqrt{3}}{2}\int dt\chi^8(t)\int\limits_{M_4}\operatorname{tr}(F^{cl})^2 \\ &= \frac{N_c}{2\sqrt{3}}\int dt\chi^8(t),\end{aligned} \tag{3.6.32}$$

where we use the baryon number is 1, i.e.,

$$N_B = \frac{1}{8\pi^2}\int\limits_{M_4}\operatorname{tr}(F^{cl})^2 = 1. \tag{3.6.33}$$

This is exactly a constraint term, but as we will see below, it just cancels out with the other two terms.

Using that $d\beta(\Phi dt, A^{cl})$ and $d\alpha_4(-iW^{-1}\dot{W}dt, A^{cl})$ behave as

$$(A^{cl})^3 \to (-igdg^{-1})^3 \propto \mathcal{P}_2 \tag{3.6.34}$$

$$F^{cl}(x) \sim 1/\xi^4 \tag{3.6.35}$$

at $\xi \to \infty$, we obtain

$$\begin{aligned}\frac{N_c}{24\pi^2}\int\limits_{\mathbb{R}\times M_4} d\beta(\Phi dt, A^{cl}) &= \frac{N_c}{24\pi^2}\frac{i}{4\sqrt{3}}\int dt\chi^8(t)\int\limits_{\partial M_4}\operatorname{tr}(-igdg^{-1})^3 \\ &= -\frac{N_c}{4\sqrt{3}}\int dt\chi^8(t)\end{aligned} \tag{3.6.36}$$

$$\frac{N_c}{24\pi^2}\int_{\mathbb{R}\times M_4} d\alpha_4(-iW^{-1}\dot{W}dt, A^{cl}) = \frac{N_c}{24\pi^2}\frac{i}{4\sqrt{3}}\int dt\chi^8(t)\int_{\partial M_4} tr(-igdg^{-1})^3$$
$$= -\frac{N_c}{4\sqrt{3}}\int dt\chi^8(t). \tag{3.6.37}$$

Here, in general, because of

$$N_B = \frac{1}{8\pi^2}\int_{M_4}\mathrm{tr}(F^{cl})^2 = \frac{-i}{24\pi^2}\int_{S^3}\mathrm{tr}(-igdg^{-1})^3, \tag{3.6.38}$$

we used

$$N_B = \frac{-i}{24\pi^2}\int_{S^3}(-igdg^{-1})^3 = \frac{1}{2}\mathcal{P}_2. \tag{3.6.39}$$

Therefore, the three terms cancel each other out and we have

$$S_{CS}[\mathcal{A}] = S_{CS}[\mathcal{A}^{cl}]. \tag{3.6.40}$$

This means that as long as we use the CS term used in Ref. [15, 16], we can conclude that the constraint term does not appear. This is a serious problem because the constraint term is important for the flavor SU(3) to regard the soliton as a baryon, as we have explained in the SU(3) Skyrme model. In Ref. [1] they point this out and have tried to solve this problem tentatively. Their new CS term leads to a well derived constraint term, but at the same time reveals some problems. We will look at this next.

3.6.2 New CS Term 1

In Skyrme mode, a constraint term emerged from the WZW term, which was defined on a 5-dimensional manifold. Accordingly, the collective coordinates were also dependent on two variables. The CS term that we have been using until now may also have a constraint term if it is constructed in correspondence with the Skyrme model. Let us now define the CS term as

$$S_{\mathrm{CS}}^{\mathrm{new\,1}} = \frac{N_c}{24\pi^2}\int_{M_6}\mathrm{tr}\mathcal{F}^3. \tag{3.6.41}$$

This new CS term leads to a constraint term. The manifold M_6 is labeled by $(t, x^M, s) = (t, \mathbf{x}, z, s)$ and is understood by a schematic like Fig. 3.6, which follows the configuration [22, 51] of WZW term in Skyrme model. We have so far

Fig. 3.6 A schematic of M_6

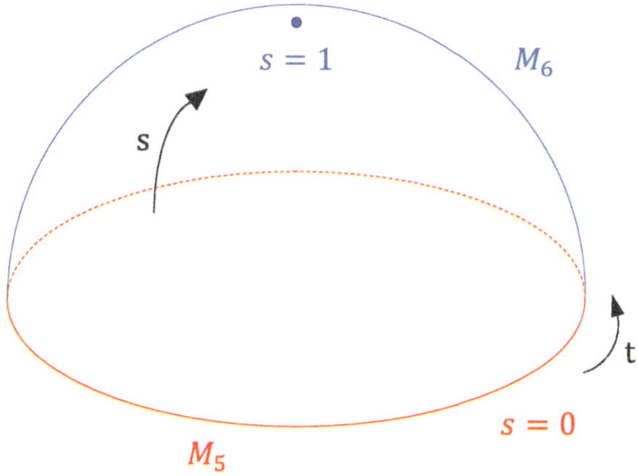

considered a 5-dimensional YM-CS theory, but to consider this new CS term, we need to introduce an additional dimension. Let us now label this axis s and introduce the corresponding gauge field component A_s. The gauge field is then written as

$$\mathcal{A}(t, x, s) = \mathcal{A}_0(t, x, s)dt + \mathcal{A}_M(t, x, s)dx^M + \mathcal{A}_s(t, x, s)ds. \qquad (3.6.42)$$

Here, the boundary of M_6 at $s = 0$ is M_5, i.e., $\partial M_6 = M_5$. Following the WZW term case, s extends in the radial direction at (t, s) and compactifies t to form disk D_2 (Fig. 3.7). Let us define the s dependence of the gauge field as

$$\mathcal{A}_0^{cl} = W(s)\mathcal{A}_0^{cl}(x, s; X^\alpha(s))W(s)^{-1}, \quad (\mathcal{A}_0^{cl}|_{s=0} = \mathcal{A}_0^{cl}(x; X^\alpha = \text{const}), \ \mathcal{A}_0^{cl}|_{s=1} = 0) \qquad (3.6.43)$$

$$\mathcal{A}_M^{cl} = W(s)\mathcal{A}_M^{cl}(x; X^\alpha(s))W(s)^{-1}, \quad (\mathcal{A}_M^{cl}|_{s=0} = \mathcal{A}_M^{cl}(x; X^\alpha = \text{const})) \qquad (3.6.44)$$

$$\mathcal{A}_s^{cl} = -iW(s)\partial_s W(s)^{-1} \qquad (3.6.45)$$

$$[\mathcal{A}_0^{cl}(x, s), g(x)] = 0. \qquad (3.6.46)$$

In this way, we can use the same solutions (3.2.8), (3.2.9), (3.2.10) and (3.2.14) as a static (time-independent) classical solution on $s = 0$, i.e., M_5.

For further explanation, let us discuss in detail the manifold we are now considering. This will also clarify the meaning of the expressions (3.6.43)–(3.6.46) that gives the restriction on the s-dependence. We defined the CS term as (3.6.41), and from

$$\text{tr}\mathcal{F}^3 = d\omega_5(\mathcal{A}), \qquad (3.6.47)$$

by Stokes' theorem, (3.6.41) seems to be equivalent to the original CS term. However, as discussed in the previous section, the two CS terms are not equivalent because the gauge field cannot be defined globally well-defined to form a field configuration with a finite baryon number, and two patches must be considered.

Fig. 3.7 A schematic of D_2

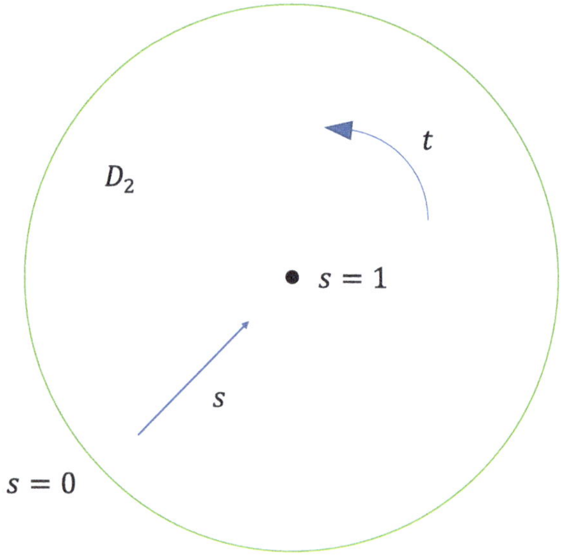

The BPST instanton solution we have adopted as the solution to the baryon number 1 is

$$
\begin{aligned}
A_M^{cl}(x) &= -if(\xi)g(x)\partial_M g(x)^{-1} \\
f(\xi) &= \frac{\xi^2}{\xi^2+\rho^2}, \quad \xi = \sqrt{(x^M - X^M)^2} \\
g(x) &= \begin{pmatrix} g^{SU(2)}(x) & 0 \\ 0 & 1_{N_f-2} \end{pmatrix}, \quad g^{SU(2)}(x) = \frac{1}{\xi}\big((z-Z)\mathbf{1}_2 + i(x^i - X^i)\tau_i\big) \\
g\partial_M g^{-1} &= \begin{cases} \frac{i}{\xi^2}\big((z-Z)\tau^i - \epsilon_{ija}(x^j - X^j)\tau^a\big) \\ -\frac{i}{\xi^2}(x^a - X^a\tau^a \end{cases},
\end{aligned}
\tag{3.6.48}
$$

which diverges at $z \to \infty$. Now, if we define

$$
\overline{A}_M^{cl}(x) = g(x)^{-1}\big(A_M^{cl} - i\partial_M\big)g(x) = -i\big(1 - f(\xi)\big)g(x)^{-1}\partial_M g(x) \tag{3.6.49}
$$

by gauge transformation, we can obtain a regular gauge field at $z \to \infty$ (it diverges at $z \to 0$).

Now the BPST instanton solution is defined on the manifold as shown in Fig. 3.8.

M_4 in $M_5 = \mathbb{R} \times M_4$ ($\mathbb{R}$ is the time axis) is covered by two patches $M_4^{(0)}$ and $M_4^{(\infty)}$ ($M_4 = M_4^{(0)} + M_4^{(\infty)}$). We define $\mathcal{A}^{cl}(x)$ on $M_4^{(0)}$ and $\overline{\mathcal{A}}^{cl}(x)$ on $M_4^{(\infty)}$, and on $B = \partial M_4^{(0)} = -\partial M_4^{(\infty)}$ where the two patches overlap, we connect the two gauge configurations by a gauge transformation. Since the time components $\hat{A}_0^{cl}$ and A_0^{cl} of

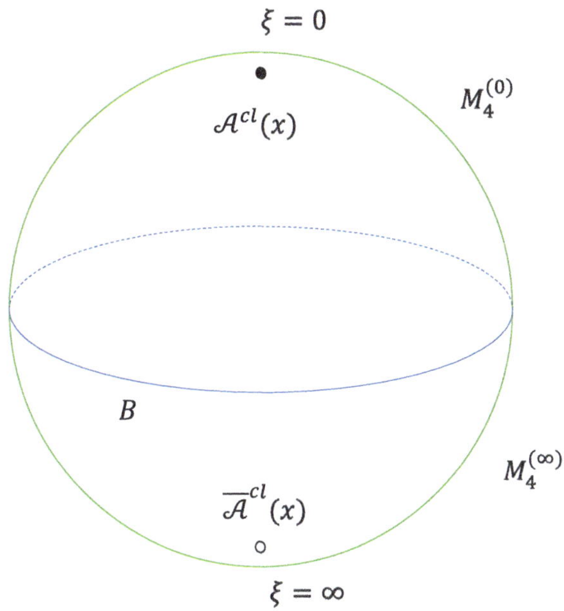

Fig. 3.8 The manifold where the BPST solution is defined

the solution are written by **1** and t_8, they are invariant to the transformation of $g(x)$ and regular in the whole region, so the same form of solution is available in the two patches. From the above discussion, we can define

$$\overline{\mathcal{A}}^{cl}(x) = (\mathcal{A}^{cl}(x))^{g(x)^{-1}} = g(x)^{-1}(\mathcal{A}^{cl}(x) - id)g(x) \tag{3.6.50}$$

as a gauge field that is regular on $M_4^{(\infty)}$, including U(1) part and time components.

Now, let us give a time dependence to the collective coordinates and extend the gauge field on M_5. On $M_5^{(0)} = \mathbb{R} \times M_4^{(0)}$, we can use

$$\mathcal{A}(t, x) = \left(\mathcal{A}^{cl}(x; X^\alpha(t)) + \Phi(t, x)dt\right)^{W(t)} \tag{3.6.51}$$

On $M_5^{(\infty)} = \mathbb{R} \times M_4^{(\infty)}$, it is reasonable to define

$$\overline{\mathcal{A}}(t, x) = \left(\overline{\mathcal{A}}^{cl}(x; X^\alpha(t)) + \overline{\Phi}(t, x)dt\right)^{W(t)} = \mathcal{A}(t, x)^{W(t)g(x;X(t))^{-1}W(t)^{-1}} \tag{3.6.52}$$

$$\begin{aligned}\overline{\Phi}(t, x) &= g(x; X(t))^{-1}\left(\Phi(t, x) - i\partial_0\right)g(x; X(t)) \\ &= -\dot{X}^N(t)\overline{\mathcal{A}}_N^{cl}(x; X^\alpha(t)) + \sum_{a=1}^{8} \chi^a(t)u^a(\xi)t_a\end{aligned} \tag{3.6.53}$$

$$\left(\mathcal{A}^{V_1V_2} = \left(\mathcal{A}^{V_2}\right)^{V_1}\right). \tag{3.6.54}$$

Let us now define a gauge field on the manifold M_6 described in Fig. 3.6. With $M_6 = D_2 \times M_4 = M_6^{(0)} + M_6^{(\infty)} (M_6^{(0/\infty)} = D_2 \times M_4^{(0/\infty)})$), we use

$$\mathcal{A}(t, x, s) = \mathcal{A}_0(t, x, s)dt + \mathcal{A}_M(t, x, s)dx^M + \mathcal{A}_s(t, x, s)ds \tag{3.6.55}$$

on $M_6^{(0)}$, as shown in (3.6.42). At $s = 0$, (3.6.41) should return to the original CS term, so the gauge field should satisfy

$$\mathcal{A}(t, x, s = 0) = \mathcal{A}(t, x). \tag{3.6.56}$$

Also, since $s = 1$ is the center of disk D_2,

$$\mathcal{O}(t, x, s = 1) = t \text{ indep.}, \quad \partial_s \mathcal{O}(t, x, s)|_{s=1} = 0, \quad (\mathcal{O} = W, X^\alpha, \Phi, \Psi) \tag{3.6.57}$$

should also be satisfied. Furthermore, we need to require

$$\mathcal{A}_0^{cl}(t, x, s = 1) = 0 \tag{3.6.58}$$

for the time component. This is necessary so that the mass formula does not change form at the classical level and corresponds to $\mathcal{A}_0^{cl}|_{s=1} = 0$ in (3.6.43).

On $M_6^{(\infty)}$, we use the gauge field

$$\begin{aligned}\overline{\mathcal{A}}(t, x, s) &= \left(\overline{\mathcal{A}}^{cl}(x, s; X^\alpha(t, s)) + \overline{\Phi}(t, x, s)dt + \overline{\Psi}(t, x, s)ds\right)^{W(t,s)} \\ &= \mathcal{A}(t, x, s)^{W(t,s)g(x;X(t,s))^{-1}W(t,s)^{-1}} \end{aligned} \tag{3.6.59}$$

$$\overline{\Phi}(t, x, s) = g(x; X(t, s))^{-1}\left(\Phi(t, x, s) - i\partial_0\right)g(x; X(t, s)) \tag{3.6.60}$$

$$\overline{\Psi}(t, x, s) = g(x; X(t, s))^{-1}\left(\Psi(t, x, s) - i\partial_s\right)g(x; X(t, s)). \tag{3.6.61}$$

We require

$$\overline{\mathcal{A}}(t, x, s = 0) = \overline{\mathcal{A}}(t, x), \quad \overline{\mathcal{A}}(t, x, s = 1) = t \text{ indep.} \tag{3.6.62}$$

in the same way as for $\mathcal{A}(t, x, s)$. Also, since $\hat{A}_0^{cl}(t, x)$ and $A_0^{cl}(t, x)$ are, as already mentioned, invariant to the transformation of $g(x)$ and regular in the whole region, we can use solutions of the same form in the two patches. In order for this property to hold when extended over M_6,

$$[\mathcal{A}_0^{cl}(x, s), g(x)] = 0 \tag{3.6.63}$$

must be satisfied, which corresponds to (3.6.46). We have now defined a gauge field on M_6.

Now prepared, let's see that a constraint term does actually emerge from this CS term. Using

$$D\delta\mathcal{A} = d\delta\mathcal{A} + i(\mathcal{A}\delta\mathcal{A} + \delta\mathcal{A}\mathcal{A}) \tag{3.6.64}$$

$$(\delta\mathcal{A})^3 = (D\delta\mathcal{A})(\delta\mathcal{A})^2 = D\mathcal{F} = 0 \tag{3.6.65}$$

$$D^2\delta\mathcal{A} = i[\mathcal{F}, \delta\mathcal{A}], \tag{3.6.66}$$

etc., with $\delta\mathcal{A} = \Phi dt + \Psi ds$, we have

$$\mathcal{F}(\mathcal{A} + \delta\mathcal{A}) = \mathcal{F}(\mathcal{A}) + D\delta\mathcal{A} + i(\delta\mathcal{A})^2 \tag{3.6.67}$$

$$\mathrm{tr}\mathcal{F}(\mathcal{A} + \delta\mathcal{A})^3 = \mathrm{tr}\mathcal{F}(\mathcal{A})^3 + 3d\mathrm{tr}\big(\delta\mathcal{A}\mathcal{F}(\mathcal{A})^2 + \delta\mathcal{A}(D\delta\mathcal{A})\mathcal{F}(\mathcal{A}\big). \tag{3.6.68}$$

Note that $S_{\mathrm{CS}}^{\mathrm{new1}}$ is gauge invariant, hence a constraint term arises like

$$\begin{aligned}
&S_{\mathrm{CS}}^{\mathrm{new}}\big[\mathcal{A} = (\mathcal{A}^{cl} + \Phi dt + \Psi ds)^W\big] - S_{\mathrm{CS}}^{\mathrm{new}}[\mathcal{A}^{cl}] \\
&= \frac{N_c}{24\pi^2}\int_{M_6} 3d\mathrm{tr}\big(\delta\mathcal{A}(\mathcal{F}^{cl})^2 + \delta\mathcal{A}(D^{cl}\delta\mathcal{A})\mathcal{F}^{cl}\big) \\
&= \frac{N_c}{8\pi^2}\int_{M_5} \mathrm{tr}\big(\Phi dt(F^{cl})^2\big) = \frac{N_c}{2\sqrt{3}}\int dt\chi^8(t)
\end{aligned} \tag{3.6.69}$$

$$\frac{N_c}{2\sqrt{3}}\chi^8(t) = \frac{N_c}{\sqrt{3}}\mathrm{tr}(-iW(t)^{-1}\dot{W}(t)t_8), \tag{3.6.70}$$

where we used

$$\delta\mathcal{A} = \Phi dt \tag{3.6.71}$$

$$\delta\mathcal{A}(D^{cl}\delta\mathcal{A}) = 0 \tag{3.6.72}$$

$$\hat{F}_{MN}^{cl} = 0 \tag{3.6.73}$$

etc. on M_5 at $S = 0$.

Here, for getting the constraint term, we used Stokes' theorem,

$$\begin{aligned}
&\frac{N_c}{24\pi^2}\int_{M_6} 3d\mathrm{tr}\big(\delta\mathcal{A}(\mathcal{F}^{cl})^2 + \delta\mathcal{A}(D^{cl}\delta\mathcal{A})\mathcal{F}^{cl}\big) \\
&= \frac{N_c}{8\pi^2}\int_{M_5} \mathrm{tr}\big(\delta\mathcal{A}(\mathcal{F}^{cl})^2 + \delta\mathcal{A}(D^{cl}\delta\mathcal{A})\mathcal{F}^{cl}\big),
\end{aligned} \tag{3.6.74}$$

which can be transformed in such a way that $\mathrm{tr}\big(\delta\mathcal{A}(\mathcal{F}^{cl})^2 + \delta\mathcal{A}(D^{cl}\delta\mathcal{A})\mathcal{F}^{cl}\big)$ is a gauge invariant quantity. Therefore, since ω_5 is not gauge-invariant, a deformation like

$$\int_{M_6} \mathrm{tr}\mathcal{F}^3 = \int_{M_5} \omega_5(\mathcal{A}) \tag{3.6.75}$$

is incorrect. The correct transformation should be

$$\int_{M_6} tr\left(\mathcal{F}^3 - (\mathcal{F}^{cl})^3\right) = \int_{M_5^{(0)}} \left(\omega_5(\mathcal{A}) - \omega_5(\mathcal{A}^{cl})\right) + \int_{M_5^{(\infty)}} \left(\omega_5(\overline{\mathcal{A}}) - \omega_5(\overline{\mathcal{A}}^{cl})\right)$$
$$+ \int_{D_2\times B} \left[\left(\omega_5(\mathcal{A}) - \omega_5(\overline{\mathcal{A}})\right) - \left(\omega_5(\mathcal{A}^{cl}) - \omega_5(\overline{\mathcal{A}}^{cl})\right)\right]. \tag{3.6.76}$$

Please refer to Ref. [1] to see that a constraint term can still be obtained in this way.

Finally, I would like to comment on the problems with $S_{\mathrm{CS}}^{\mathrm{new}1}$. First, since $\mathrm{tr}\mathcal{F}^3$ is a gauge invariant, the chiral anomaly is not reproduced from this CS term, which was well reproduced by the original CS term. In the Sakai-Sugimoto model, $Z \to \pm\infty$ is the boundary, as shown in Fig. 3.5. That is, $\partial M_5 = M_4^{-\infty} \cup M_4^{+\infty}$. However, $\partial(\partial M) = 0$ is satisfied for any manifold M in general. If $\partial M_6 = M_5$, then $\partial(\partial M) = 0$ must hold, which is a contradiction. Also, the origin of the new introduced dimension has not been clear. To solve these problems, another New CS term was considered in Ref. [2]. In the following, this CS term will be explained.

3.6.3 New CS Term 2

In this section, we use the anti-Hermitian $U(N_f)$ gauge field used in Sect. 3.6.1. The following discussion can be used not only for BPST instanton solutions.

To solve the problems described in the previous section, the following CS term was proposed;

$$S_{\mathrm{CS}}^{\mathrm{new}2} = C\Bigg(\int_{M_5^-} \omega_5(A_-) + \int_{M_5^+} \omega_5(A_+)$$
$$+ \frac{1}{10}\int_{N_5^{(0)}} \mathrm{tr}\left((\tilde{h}d\tilde{h}^{-1})^5\right) + \int_{M_4^{(0)}} a_4(dh^{-1}h, A_-)\Bigg), \tag{3.6.77}$$

where M_5^-, M_5^+, $M_4^{(0)}$ is defined in Fig. 3.5. $N_5^{(0)}$ were defined in Fig. 3.9 (This figure is only as an help for understanding, and the following figures are also not so much concerned with a rigorous). $N_5^{(0)}$ is a manifold such that it satisfies $\partial N_5^{(0)} = M_4^{(0)}$. We define $\tilde{h} \in U(N_f)$ on $N_5^{(0)}$ which satisfies $\tilde{h}|_{\partial N_5^{(0)}} = h$. Since we are now using an anti-Hermitian gauge field, (3.6.24) is rewritten as

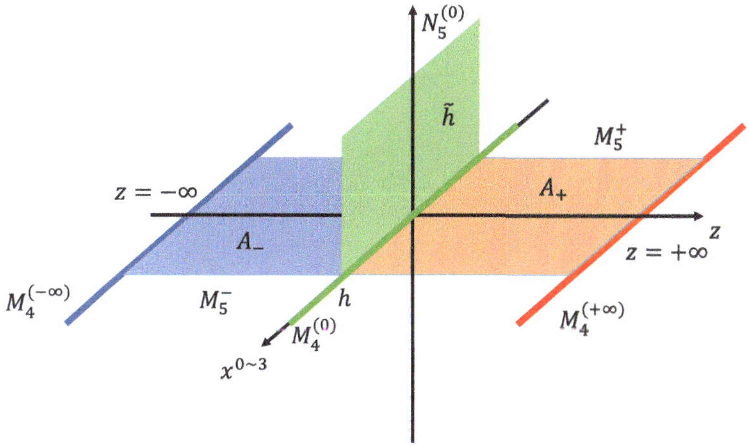

Fig. 3.9 Manifolds on which $S_{\mathrm{CS}}^{\mathrm{new2}}$ are defined

$$\alpha_4(V, A) = \frac{1}{2}\mathrm{tr}\Big(V(A^3 - AF - FA) + \frac{1}{2}VAVA + V^3A\Big) \tag{3.6.78}$$

$$= -\frac{1}{2}\mathrm{tr}\Big(V(AdA + dAA + A^3) - \frac{1}{2}VAVA - V^3A\Big). \tag{3.6.79}$$

This CS term seems artificial but we can immediately find that it is determined quite uniquely from several physical requirements.

We now discuss the assumptions for defining this CS term. We need to assume the existence of $N_5^{(0)}$ and the field $\tilde{h} \in U(N_f)$ defined on it, as described above. This corresponds to the fact that in the discussion of $S_{\mathrm{CS}}^{\mathrm{new1}}$, we assumed the existence of a new dimension s and a gauge field $\mathcal{A}_s$ defined on it.

We will see that $S_{\mathrm{CS}}^{\mathrm{new2}}$ is appropriate for the CS term we are looking for and discuss the meaning of each term in the Eq. (3.6.77). $S_{\mathrm{CS}}^{\mathrm{new2}}$ satisfies the following properties.

1. When h is topologically trivial (baryon number is 0), $S_{\mathrm{CS}}^{\mathrm{new2}}$ returns to the original CS term, (3.6.2).
2. Under a gauge transformation (3.6.16) satisfying $g_\pm|_{z\to\pm\infty} \to 1$, $S_{\mathrm{CS}}^{\mathrm{new2}}$ is invariant (shift of $2\pi\mathbb{Z}$).
3. If we identify $\hat{A}_\pm = A_\pm|_{z\to\pm\infty}$ and $\hat{g}_\pm = e^{-\hat{\Lambda}_\pm} = g_\pm|_{z\to\pm\infty}$, it reproduces the QCD chiral anomaly (3.6.5).
4. If a manifold M_6 (Fig. 3.10) exists such that M_5 has no boundary and satisfies $(M_4^{(\pm\infty)} = \emptyset)$, $\partial M_6 = M_5$, $M_6 = M_6^+ \cup M_6^-$, $M_6^+ \cap M_6^- = N_5^{(0)} \times (-\epsilon, \epsilon)$, and $\partial M_6^\pm = M_5^\pm \cup (\pm N_5^{(0)})$, then $S_{\mathrm{CS}}^{\mathrm{new2}}$ becomes (3.6.41).

The following explanation shows that the CS term that satisfies these physically required properties is determined quite uniquely to be (3.6.77).

Let us look at these properties of $S_{\mathrm{CS}}^{\mathrm{new2}}$ one by one.

First consider the Property 1. When h is topologically trivial, that is, when h can be transformed to $h = 1$ by continuous transformation, we can define $\tilde{h} \in U(N_f)$

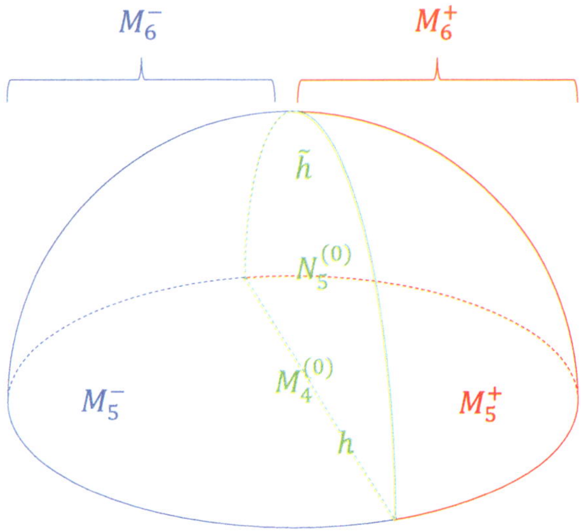

Fig. 3.10 The manifold M6 on which $S_{\rm CS}^{\rm new1}$ is defined

on M_5^- that satisfies $\tilde{h} = h$ on $M_5^- \cup M_5^+$ and $\tilde{h}|_{z\to-\infty} \to 1$ on the boundary. Using this $\tilde{h}$, we see that we can define a globally well-defined 1-form A over M_5 as

$$A = \begin{cases} A_-^{\tilde{h}}, & (\text{on } M_5^-) \\ A_+, & (\text{on } M_5^+) \end{cases} . \tag{3.6.80}$$

Now if we define $N_5^{(0)} = M_5^- \cup N_5^{-\infty}$ ($\partial N_5^{-\infty} = M_4^{-\infty}$, see also Fig. 3.11 for details), we find that with $\tilde{h}|_{N_5^{(-\infty)}} = 1$, $\tilde{h}$ is a function on $N_5^{(0)}$ that returns to the original CS term, like

$$\begin{aligned} S_{\rm CS}^{\rm nem2} =& C\Big(\int_{M_5^-} \omega_5(A_-) + \int_{M_5^+} \omega_5(A_-) \\ &+ \int_{M_5^-} \Big[\frac{1}{10} \mathrm{tr}\big((\tilde{h} d\tilde{h}^{-1})^5\big) + d\alpha_4\big(d\tilde{h}^{-1}\tilde{h}, A_-\big) \Big] \Big) \\ =& C\Big(\int_{M_5^-} \omega_5(A_-^{\tilde{h}}) + \int_{M_5^+} \omega_5(A_+) \Big) \\ =& C \int_{M_5} \omega_5(A). \end{aligned} \tag{3.6.81}$$

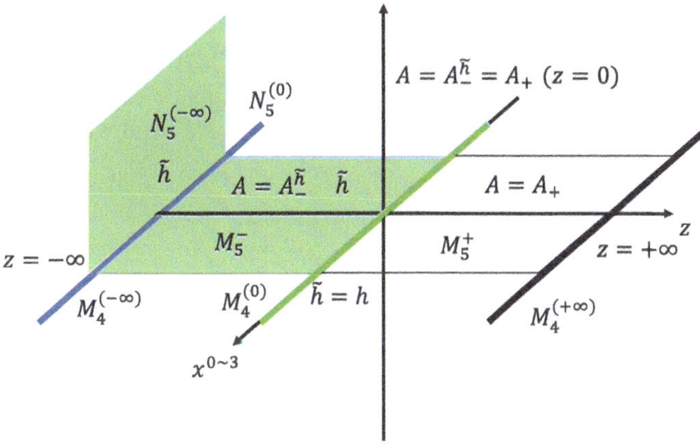

Fig. 3.11 The manifolds for which property 1 holds

Here, we used the formula

$$\omega_5(A^g) = \omega_5(A) + \frac{1}{10}\mathrm{tr}\big((gdg^{-1})^5\big) + d\alpha_4\big(dg^{-1}g,\, A\big), \tag{3.6.82}$$

which corresponds to the (3.6.22) formula when using an anti-Hermitian gauge field, and the fact that the fourth term of (3.6.77) can be written as

$$\begin{aligned}\int_{M_4^{(0)}} \alpha_4(dh^{-1}h,\, A_-) = \int_{N_5^{(0)}} d\alpha_4 &= \int_{M_5^-} d\alpha_4 + \int_{N_5^{(-\infty)}} d\alpha_4(dh^{-1}h|_{z\to-\infty},\, A_-) \\ &= \int_{M_5^-} d\alpha_4(dh^{-1}h,\, A_-)\end{aligned} \tag{3.6.83}$$

in the present case.

Next consider the Property 2. Under the gauge transformation,

$$A_\pm \to A_\pm^{g_\pm} = g_\pm A_\pm g_\pm^{-1} + g_\pm dg_\pm^{-1}, \quad h \to g_+ h g_-^{-1}, \quad (g_\pm|_{z\to\pm\infty} \to 1), \tag{3.6.84}$$

$S_{\mathrm{CS}}^{\mathrm{new2}}$ can be written as

$$\begin{aligned}S_{\mathrm{CS}}^{\mathrm{new2}} \to C\Big(&\int_{M_5^-} \omega_5(A_-^{g_-}) + \int_{M_5^+} \omega_5(A_+^{g_+}) \\ &+ \frac{1}{10}\int_{N_5^{(0)}} \mathrm{tr}\big((\tilde{h}'d\tilde{h}'^{-1})^5\big) + \int_{M_4^{(0)}} \alpha_4\big(d\tilde{h}'^{-1}\tilde{h}',\, A_-^{g_-}\big)\Big),\end{aligned} \tag{3.6.85}$$

where $h' = g_+ h g_-^{-1}$ and $\tilde{h}'$ are functions that take values in $U(N_f)$ on $M_5^- \cap M_5^+$ and $N_5^{(0)}$ respectively, which satisfy $\tilde{h}'|_{\partial N_5^{(0)}} = h'|_{z=0}$. At the boundary, $g_\pm|_{z=0}$ is topologically trivial, because of $g_\pm|_{z\to\pm\infty} \to 1$. In such a case, we can define a function $\tilde{g}_\pm$ such that $\tilde{g}_\pm|_{\partial N_5^{(0)}} = g_\pm|_{z=0}$ is satisfied on $N_5^{(0)}$ and write $\tilde{h}' = \tilde{g}_+ \tilde{h} \tilde{g}_-^{-1}$. Using formula

$$\alpha_4(V, \pm V) = 0 \tag{3.6.86}$$

$$\alpha_4(dgg^{-1}, A^g) = -\alpha_4(dg^{-1}g, A) \tag{3.6.87}$$

$$\begin{aligned} \alpha_4(d(gh)(gh)^{-1} A^g) &= \alpha_4(g(H - G)g^{-1}, A^g) \\ &= \alpha_4(H, A) - \alpha_4(G, A) \\ &\quad - \frac{1}{2}\mathrm{tr}\Big(G^3 H + G H^3 - \frac{1}{2} GHGH\Big) \\ &\quad + \frac{1}{2} d\mathrm{tr}\big((H - G)(AG - GA)\big) \end{aligned} \tag{3.6.88}$$

$$\begin{aligned} \alpha_4(d(gh)^{-1}(gh), A) &= \alpha_4(dh^{-1}h, A) + \alpha_4(G, A^h) \\ &\quad + \frac{1}{2}\mathrm{tr}\Big(G^3 H + G H^3 - \frac{1}{2} GHGH\Big) \\ &\quad - \frac{1}{2} d\mathrm{tr}\big((H - G)(A^h G - G A^h)\big) \end{aligned} \tag{3.6.89}$$

$$\begin{aligned} \mathrm{tr}((U^{-1}dU)^5) &= -\mathrm{tr}(G^5) + \mathrm{tr}(H^5) \\ &\quad + 5d\mathrm{tr}\Big(G^3 H + G H^3 - \frac{1}{2} GHGH\Big) \end{aligned} \tag{3.6.90}$$

$$(G = dg^{-1}g, \quad H = dhh^{-1}, \quad U = gh)$$

and (3.6.22), etc., $S_{\mathrm{CS}}^{\mathrm{new2}}$ can be gauge transformed as follows,

$$\begin{aligned} S_{\mathrm{CS}}^{\mathrm{new2}} \to C\Big(&\int_{M_5^-} \omega_5(A_-) + \int_{M_5^+} \omega_5(A_+) \\ &+ \frac{1}{10}\int_{N_5^{(0)}} \mathrm{tr}\big((\tilde{h} d\tilde{h}^{-1})^5\big) + \int_{M_4^{(0)}} \alpha_4\big(dh^{-1}h, A_-\big)\Big) \\ &+ \frac{C}{10}\Big(\int_{M_5^+} \mathrm{tr}(G_+^5) + \int_{N_5^{(0)}} \mathrm{tr}(\tilde{G}_+^5)\Big) \\ &+ \frac{C}{10}\Big(\int_{M_5^-} \mathrm{tr}(G_-^5) + \int_{N_5^{(0)}} \mathrm{tr}(\tilde{G}_-^5)\Big), \end{aligned} \tag{3.6.91}$$

with $G_\pm = dg_\pm^{-1} g_\pm$ and $\tilde{G}_\pm = d\tilde{g}_\pm^{-1}\tilde{g}_\pm$. The first and second lines of this equation are $S_{\mathrm{CS}}^{\mathrm{new2}}$ (3.6.77), and the third and fourth lines take the value $2\pi\mathbb{Z}$. Thus, we can see that the third and fourth terms in (3.6.77) were introduced to just cancel out the changes in the first and second terms by this gauge transformation. We can also see that Property 1, which reduces to the original CS term when h is trivial, is also ensured by the third and fourth terms.

Next, let us consider the Property 3. Under the infinitesimal gauge transformation $\hat{g}_\pm \simeq 1 - \Lambda_\pm$, from the first and second terms of $S_{\rm CS}^{\rm new2}$ (3.6.77), as shown in (3.6.5), the QCD chiral anomaly emerges. The third and fourth terms of (3.6.77) also change with this gauge transformation. However, in the vicinity of $z = 0$ where these terms exist, the gauge transformation $g_\pm$ we are now considering can be regarded as trivial. Therefore, as explained in Property 2, the change of this two terms is equal to $2\pi\mathbb{Z}$. From the above, the chiral anomaly of QCD is reproduced. It can be seen that the form of the first and second terms in (3.6.77) are required to produce the chiral anomaly.

Finally, let us look at the Property 4. When M_5 has no boundary and there exists a manifold M_6 such that Fig. 3.10, we find that $\partial M_6^\pm = M_5^\mp \cup (\pm N_5^{(0)})$, then using Stokes' theorem, $S_{\rm CS}^{\rm new1}$ is transformed to

$$\begin{aligned} S_{\rm CS}^{\rm new1} =& C\Big(\int_{M_6^-} d\omega_5(A_-) + \int_{M_6^+} d\omega_5(A_+) \Big) \\ =& C\Big(\int_{M_5^-} \omega_5(A_-) + \int_{M_5^+} \omega_5(A_+) \\ &+ \int_{N_5^{(0)}} \big(\omega_5(A_+) - \omega_5(A_-)\big)\Big). \end{aligned} \tag{3.6.92}$$

Since $A_+ = A_-^{\tilde{h}}$ on $M_6^- \cap M_6^+ = N_5^{(0)} \times (-\epsilon, +\epsilon)$, and using (3.6.22), we find

$$\begin{aligned} &\int_{N_5^{(0)}} \big(\omega_5(A_+) - \omega_5(A_-)\big) \\ &= \int_{N_5^{(0)}} \frac{1}{10}{\rm tr}\big((\tilde{h}d\tilde{h}^{-1})^5\big) + \int_{\partial N_5^{(0)}} \alpha_4\big(d\tilde{h}^{-1}\tilde{h}, A_-\big), \end{aligned} \tag{3.6.93}$$

so in such a case, $S_{\rm CS}^{\rm new2}$ is equivalent to $S_{\rm CS}^{\rm new1}$. Again, we understand the importance of the third and fourth terms in (3.6.77). It should be emphasized that the Property 4 suggests that $S_{\rm CS}^{\rm new2}$ has the potential to correctly lead to a constraint term. As we will explain later, we find that $S_{\rm CS}^{\rm new2}$ involves $S_{\rm CS}^{\rm new1}$, not only in the current special case, so this CS term leads to the constraint term correctly, as we had expected.

From the above explanation, it is understood that our desired CS term is quite uniquely determined in the form $S_{\rm CS}^{\rm new2}$ (3.6.77).

We show some alternative expressions for $S_{\rm CS}^{\rm new2}$ (3.6.77). From one of these expressions, we can conclude that $S_{\rm CS}^{\rm new2}$ leads to a constraint term. First, let us define a new, globally well-defined gauge field A that can take non-zero baryon numbers. Such a gauge field can be defined as

$$A = A_{\pm}^{h_{\pm}}, \quad (\text{on } M_5^{\pm}) \tag{3.6.94}$$

$$h_+ h h_-^{-1} = 1, \quad A_+^{h_+} = A_-^{h_-}, \quad (\text{on } M_5^- \cap M_5^+). \tag{3.6.95}$$

The A at the boundary is not exactly equal to the external field $\hat{A}_{\pm}$ and becomes

$$A|_{z\to\pm\infty} = \hat{A}_{\pm}^{\hat{h}_{\pm}}, \quad (\hat{h}_{\pm} = h_{\pm}|_{z\to\pm\infty}). \tag{3.6.96}$$

The field configuration of a global gauge field defined in this way should be specified by $(A, \hat{h}_{\pm})$. That is, we need to physically distinguish between $(A, \hat{h}_{\pm})$ and $(A, \hat{h}'_{\pm})$. When the external field $\hat{A}_{\pm}$ is zero, the baryon number is defined to be

$$N_B = -\frac{1}{24\pi^2}\int_{S^3} \left(\mathrm{tr}((\hat{h}_+ d\hat{h}_+^{-1})^3) - \mathrm{tr}((\hat{h}_- d\hat{h}_-^{-1})^3)\right) \tag{3.6.97}$$

with $A|_{z\to\pm\infty} = \hat{h}_{\pm} d\hat{h}_{\pm}^{-1}$. If $\hat{h}_{\pm}$ is topologically nontrivial, it gives a finite baryon number.

If we now consider a gauge transformation like

$$A \to A^g, \quad \hat{h}_{\pm} \to (g\hat{h}_{\pm})|_{z\to\pm\infty}, \tag{3.6.98}$$

we see that this gauge transformation does not act on the external field $\hat{A}_{\pm}$, i.e.,

$$\hat{A}_{\pm} \to \hat{A}_{\pm}, \quad \left(\text{under } (A, \hat{h}_{\pm}) \to (A^g, g\hat{h}_{\pm})\right). \tag{3.6.99}$$

For the chiral transformation, by defining $(\hat{g}_-, \hat{g}_+) \in U(N_f)_L \times U(N_f)_R$ as

$$\hat{g}_{\pm} = (\hat{h}_{\pm}^{-1} g \hat{h}_{\pm})|_{z\to\pm\infty} \tag{3.6.100}$$

and considering the gauge transformation

$$A \to A^g, \quad \hat{h}_{\pm} \to \hat{h}_{\pm}, \tag{3.6.101}$$

we obtain

$$\hat{A}_{\pm} \to \hat{A}_{\pm}^{\hat{g}_{\pm}}, \quad \left(\text{under } (A, \hat{h}) \to (A^g, \hat{h})\right). \tag{3.6.102}$$

Following the same calculation as when we showed the Property 2, (3.6.77) is rewritten in the form

$$S_{\rm CS}^{\rm new2} = C\Big(\int_{M_5} \omega_5(A) + \int_{N_5^{(+\infty)}} \frac{1}{10}{\rm tr}\big((h_+^{-1}dh_+)^5\big) + \int_{M_4^{(+\infty)}} \alpha_4\big(d\hat{h}_+\hat{h}_+^{-1}, A\big) - \int_{N_5^{(-\infty)}} \frac{1}{10}{\rm tr}\big((h_-^{-1}dh_-)^5\big) - \int_{M_4^{(-\infty)}} \alpha_4\big(d\hat{h}_-\hat{h}_-^{-1}, A\big)\Big), \tag{3.6.103}$$

using the globally well-defined gauge field A defined by (3.6.94). Here, $N_5^{(\pm\infty)}$ is a five-dimensional manifold satisfying $\partial N_5^{(\pm\infty)} = M_5^{(\pm\infty)}$ and $h_\pm$ is a function which takes the value of $U(N_f)$ on $N_5^{(\pm\infty)}$ satisfying $h_\pm|_{\partial N_5^{(\pm\infty)}} = \hat{h}_\pm$. Moreover, using (3.6.87), it can be transformed to

$$S_{\rm CS}^{\rm new2} = C\Big(\int_{M_5} \omega_5(A) + \int_{N_5^{(+\infty)}} \frac{1}{10}{\rm tr}\big((h_+^{-1}dh_+)^5\big) - \int_{M_4^{(+\infty)}} \alpha_4\big(d\hat{h}_+^{-1}\hat{h}_+, \hat{A}_+\big) - \int_{N_5^{(-\infty)}} \frac{1}{10}{\rm tr}\big((h_-^{-1}dh_-)^5\big) + \int_{M_4^{(-\infty)}} \alpha_4\big(d\hat{h}_-^{-1}\hat{h}_-, A_-\big)\Big). \tag{3.6.104}$$

From this expression, we see that $S_{\rm CS}^{\rm new2}$ is equivalent to the original CS term (3.6.2) when $\hat{A}_\pm = 0$ with $N_f = 2$.

Furthermore, we define the manifold N_5 as

$$\begin{aligned} N_5 &= N_5^{(+\infty)} \cup \big(-N_5^{(-\infty)}\big) \\ \partial N_5 &= \partial M_5 = M_4^{(+\infty)} \cup \big(-M_4^{(-\infty)}\big) \end{aligned} \tag{3.6.105}$$

and the function h of $U(N_f)$ on N_5 as

$$\hat{h}_\pm = h|_{M_4^{(\pm\infty)}} \tag{3.6.106}$$

to get a more simplified form

$$\begin{aligned} S_{\rm CS}^{\rm new2} &= C\Big(\int_{M_5} \omega_5(A) + \int_{N_5} \frac{1}{10}{\rm tr}\big((h^{-1}dh)^5\big) + \int_{\partial M_5} \alpha_4\big(dhh^{-1}, A\big)\Big) \\ &= C\Big(\int_{M_5} \omega_5(A) + \int_{N_5} \frac{1}{10}{\rm tr}\big((h^{-1}dh)^5\big) + \int_{\partial M_5} \alpha_4\big(dh^{-1}h, \hat{A}\big)\Big), \end{aligned} \tag{3.6.107}$$

where $\hat{A}$ is represented by

$$\hat{A}_\pm = \hat{A}|_{M_4^{(\pm\infty)}} \tag{3.6.108}$$

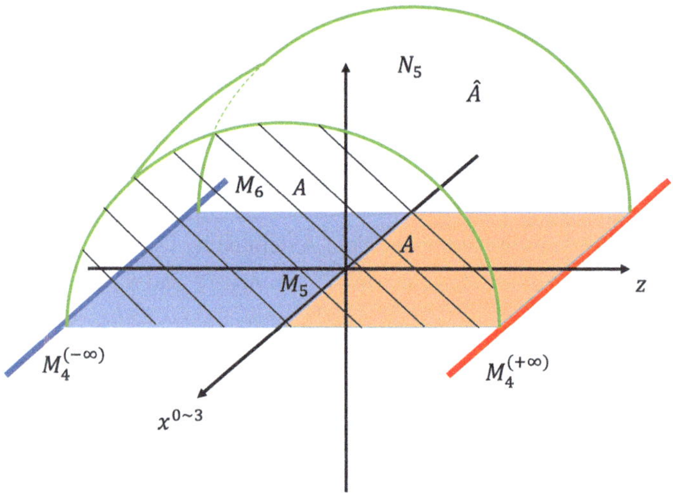

Fig. 3.12 M_6 in $S_{\rm CS}^{\rm new2}$ ($N_5^{-\infty}$ and $N_5^{+\infty}$ are compactified)

and the (3.6.95) formula is represented by

$$A|_{\partial M_5} = \hat{A}^h. \tag{3.6.109}$$

The gauge transformation (3.6.98), which does not act on the boundary, is

$$A \to A^g, \quad h \to gh, \quad \hat{A} \to \hat{A}. \tag{3.6.110}$$

The chiral transformation (3.6.101) is represented as

$$A \to A^g, \quad h \to h, \quad \hat{A} \to \hat{A}^{\hat{g}} \tag{3.6.111}$$

$$\hat{g} = (h^{-1}gh)|_{\partial M_5}, \quad \hat{g}_\pm = \hat{g}|_{M_4^{(\pm\infty)}}. \tag{3.6.112}$$

It is easy to see that the Properties 2 and 3 are satisfied, respectively.

Now we can derive the constraint terms, while solving the problems of Ref. [1]. Define the manifold M_6 as such that it satisfies

$$\partial M_6 = M_5 \cup (-N_5) \tag{3.6.113}$$

(see Fig. 3.12). Since $M_5 \cup (-N_5)$ has no boundary, such a manifold M_6 is possible. Also, if we focus only on M_5, this manifold itself can have a boundary, so a manifold like (3.6.1) is possible. From the above, we find

$$\int_{M_6} tr(F^3) = \int_{M_5} \omega_5(A) - \int_{N_5} \omega_5(A), \tag{3.6.114}$$

and if we define the external field on N_5 as $\hat{A} = A^{h^{-1}}$ and note that $\partial N_5 = \partial M_5$, we find

$$\int_{N_5} \omega_5(\hat{A}) = \int_{N_5} \left(\omega_5(A) + \frac{1}{10}\mathrm{tr}\left((h^{-1}dh)^5\right)\right) + \int_{\partial N_5} \alpha_4\left(dhh^{-1}, A\right), \quad (3.6.115)$$

so we obtain a new expression for

$$S_{\mathrm{CS}}^{\mathrm{new2}} = C\left(\int_{M_6} \mathrm{tr}(F^3) + \int_{N_5} \omega_5(\hat{A})\right) \quad (3.6.116)$$

with a more explicit connection to $S_{\mathrm{CS}}^{\mathrm{new1}}$. From this, we observe that from the first term of (3.6.116), we derive a constraint term. The chiral anomaly is also correctly derived, as explained using another expression.

3.7 Appendix: Details of the Calculation

3.7.1 *Equation (3.2.25)*

It is almost the same as Appendix 2.7.6. With

$$\overline{A}_M^{cl}(x) = g(x)^{-1}\left(A^{cl}(x) - i\partial_M\right)g(x) = -i(1 - f(\xi))g(x)^{-1}\partial_M g(x) \quad (3.7.1)$$

$$\overline{\Phi}_{SU(3)}(t, x) = g(x; X(t))^{-1}\Phi_{SU(3)}(t, x)g(x; X(t)) \quad (3.7.2)$$

The only difference is that the solution of

$$\overline{D}_M^{cl}\overline{D}_M^{cl}\overline{\Phi}_{SU(3)} = 0. \quad (3.7.3)$$

we obtain as follows;

$$\overline{\Phi}_{SU(3)} = \sum_{a=1}^{8} \chi^a(t)u^a(\xi)t_a \quad (3.7.4)$$

$$\frac{1}{\xi^3}\frac{d}{d\xi}\left(\xi^3\frac{d}{d\xi}u^a(\xi)\right) = C_a\frac{(1 - f(\xi))^2}{\xi^2}u^a(\xi) \quad (3.7.5)$$

$$C_a = \begin{cases} 8, & (a = 1, 2, 3) \\ 3, & (a = 4, 5, 6, 7) \\ 0, & (a = 8) \end{cases} . \quad (3.7.6)$$

3.7.2 Equation (3.3.11)

From

$$\mathbf{L}_{[\xi m,\xi_n]} s = f_{mnp} \mathbf{L}_{\xi_p} s \tag{3.7.7}$$

we derive

$$[\xi_m, \xi_n]^{\hat{\alpha}} = f_{mnp} \xi_p^{\hat{\alpha}}. \tag{3.7.8}$$

We have

$$\begin{aligned}\mathbf{L}_{[\xi m,\xi_n]} s &= \xi_m^{\hat{\alpha}} \partial_{\hat{\alpha}} (\xi_n^{\hat{\beta}} \partial_{\hat{\beta}}) s - \xi_n^{\hat{\alpha}} \partial_{\hat{\alpha}} (\xi_m^{\hat{\beta}} \partial_{\hat{\beta}}) s = (\xi_m^{\hat{\alpha}} \partial_{\hat{\alpha}} \xi_n^{\hat{\beta}} - \xi_n^{\hat{\alpha}} \partial_{\hat{\alpha}} \xi_m^{\hat{\beta}}) \partial_{\hat{\beta}} s = [\xi_m, \xi_n]^{\hat{\beta}} \partial_{\hat{\beta}} s \\ &= f_{mnp} \mathbf{L}_{\xi_p} s = f_{mnp} \xi_p^{\hat{\beta}} \partial_{\hat{\beta}} s \end{aligned} \tag{3.7.9}$$

for the arbitrary $\partial_{\hat{\beta}} s$,

$$[\xi_m, \xi_n]^{\hat{\alpha}} = f_{mnp} \xi_p^{\hat{\alpha}} \tag{3.7.10}$$

hold.

3.7.3 Equation (3.3.15)

First we prove

$$(\mathbf{L}_{\xi_m} \mathbf{L}_{\xi_n} - \mathbf{L}_{\xi_n} \mathbf{L}_{\xi_m}) A_\mu = L_\eta A_\mu \tag{3.7.11}$$

$$\eta^\mu = [\xi_m, \xi_n]^\mu = \xi_m^\rho \partial_\rho \xi_n^\mu - \xi_n^\rho \partial \rho \xi_m^\mu \tag{3.7.12}$$

The left-hand side expand

$$\begin{aligned}&(\mathbf{L}_{\xi_m} \mathbf{L}_{\xi_n} - \mathbf{L}_{\xi_n} \mathbf{L}_{\xi_m}) A_\mu \\ &= \xi_m^\rho (\partial_\rho \xi_n^\nu)(\partial_\nu \omega_\mu) + \xi_m^\rho \xi_n^\nu (\partial_\rho \partial_\nu \omega_\mu) + \xi_m^\rho (\partial_\rho \partial_\nu \xi_n^\nu) \omega_\nu \\ &\quad + \xi_m^\rho (\partial_\mu \xi_n^\nu)(\partial_\rho \omega_\nu) + (\partial_\mu \xi_m^\rho) \xi_n^\nu (\partial_\nu \omega_\rho) + (\partial_\mu \xi_m^\rho)(\partial_\rho \xi_n^\nu) \omega_\nu \\ &\quad - \xi_n^\rho (\partial_\rho \xi_m^\nu)(\partial_\nu \omega_\mu) - \xi_n^\rho \xi_m^\nu (\partial_\rho \partial_\nu \omega_\mu) - \xi_n^\rho (\partial_\rho \partial_\nu \xi_m^\nu) \omega_\nu \\ &\quad - \xi_n^\rho (\partial_\mu \xi_m^\nu)(\partial_\rho \omega_\nu) - (\partial_\mu \xi_n^\rho) \xi_m^\nu (\partial_\nu \omega_\rho) - (\partial_\mu \xi_n^\rho)(\partial_\rho \xi_m^\nu) \omega_\nu, \end{aligned} \tag{3.7.13}$$

Since the second and eighth, fourth and eleventh, and fifth and tenth terms cancel each other out, we calculate

$$
\begin{aligned}
&(\mathbf{L}_{\xi_m}\mathbf{L}_{\xi_n} - \mathbf{L}_{\xi_n}\mathbf{L}_{\xi_m})A_\mu \\
&= \xi_m^\rho(\partial_\rho\xi_n^\nu)(\partial_\nu\omega_\mu) + \xi_m^\rho(\partial_\rho\partial_\mu\xi_n^\nu)\omega_\nu + (\partial_\mu\xi_m^\rho)(\partial_\rho\xi_n^\nu)\omega_\nu \\
&\quad - \xi_n^\rho(\partial_\rho\xi_m^\nu)(\partial_\nu\omega_\mu) - \xi_n^\rho(\partial_\rho\partial_\nu\xi_m^\nu)\omega_\nu - (\partial_\mu\xi_n^\rho)(\partial_\rho\xi_m^\nu)\omega_\nu
\end{aligned} \tag{3.7.14}
$$

On the other hand, since the right-hand side is calculated as

$$
\begin{aligned}
L_\eta A_\mu &= (\xi_m^\mu\partial_\mu\xi_n^\nu - \xi_n^\mu\partial_\mu\xi_m^\nu)\partial_\nu\omega_\nu + \partial_\rho(\xi_m^\mu\partial_\mu\xi_n^\nu - \xi_n^\mu\partial_\mu\xi_m^\nu)\omega_\nu \\
&= \xi_m^\mu(\partial_\mu\xi_n^\nu)(\partial_\nu\omega_\rho) - \xi_n^\mu(\partial_\mu\xi_m^\nu)(\partial_\nu\omega_\rho) + (\partial_\rho\xi_m^\mu)(\partial_\mu\xi_n^\nu)\omega_\nu \\
&\quad + \xi_m^\mu(\partial_\rho\partial_\mu\xi_n^\nu)\omega_\nu - (\partial_\rho\xi_n^\mu)(\partial_\mu\xi_m^\nu)\omega_\nu - \xi_n^\mu(\partial_\rho\partial_\mu\xi_m^\nu)\omega_\nu,
\end{aligned} \tag{3.7.15}
$$

By comparing both sides, we can see that they are equal. We have now proved this for the case of 1-form $A_\mu dx^\mu$, but this relation holds for general n-forms.

Next, we prove

$$
\mathbf{L}_{\xi_m}(D_\mu W_n) - \mathbf{L}_{\xi_n}(D_\mu W_m) = D_\mu(\mathbf{L}_{\xi_m}W_n - \mathbf{L}_{\xi_n}W_m - [W_m, W_n]) \tag{3.7.16}
$$

The left-hand side is calculated as

$$
\begin{aligned}
&\mathbf{L}_{\xi_m}(D_\mu W_n) - \mathbf{L}_{\xi_n}(D_\mu W_m) \\
&= (\partial_\mu\xi_m^\rho)(\partial_\rho W_n + [A_\rho, W_n]) + \xi_m^\rho\partial_\rho(\partial_\mu W_n + [A_\mu, W_n]) \\
&\quad - (\partial_\mu\xi_n^\rho)(\partial_\rho W_m + [A_\rho, W_m]) - \xi_n^\rho\partial_\rho(\partial_\mu W_m + [A_\mu, W_m]) \\
&= (\partial_\mu\xi_m^\rho)(\partial_\rho W_n) + (\partial_\mu\xi_m^\rho)[A_\rho, W_n] + \xi_m^\rho(\partial_\rho\partial_\mu W_n) + \xi_m^\rho\partial_\rho([A_\mu, W_n]) \\
&\quad - (\partial_\mu\xi_n^\rho)(\partial_\rho W_m) - (\partial_\mu\xi_n^\rho)[A_\rho, W_m] - \xi_n^\rho(\partial_\rho\partial_\mu W_m) - \xi_n^\rho\partial_\rho([A_\mu, W_m])
\end{aligned} \tag{3.7.17}
$$

The right-hand side becomes

$$
\begin{aligned}
&D_\mu(\mathbf{L}_{\xi_m}W_n - \mathbf{L}_{\xi_n}W_m - [W_m, W_n]) \\
&= \partial_\mu(\xi_m^\rho\partial_\rho W_n - \xi_\rho^\rho W_m + [W_m, W_n]) + [A_\mu, \xi_m^\rho\partial_\rho W_n] - [A_\mu, \xi_n^\rho\partial_\rho W_m] + [A_\mu, [W_m, W_n]] \\
&= (\partial_\mu\xi_m^\rho)(\partial_\rho W_n) + \xi_m^\rho(\partial_\mu\partial_\rho W_n) - (\partial_\mu\xi_n^\rho)(\partial_\rho W_m) - \xi_n^\rho(\partial_\mu\partial_\rho W_m) \\
&\quad + \xi_m^\rho\partial_\rho([A_\mu, W_n]) - [\xi_m^\rho\partial_\rho A_\mu, W_n] - \xi_n^\rho\partial_\rho([A_\mu, W_m]) + [\xi_n^\rho\partial_\rho A_\mu, W_m] + D_\mu([W_m, W_n]) \\
&= (\partial_\mu\xi_m^\rho)(\partial_\rho W_n) + \xi_m^\rho(\partial_\mu\partial_\rho W_n) - (\partial_\mu\xi_n^\rho)(\partial_\rho W_m) - \xi_n^\rho(\partial_\mu\partial_\rho W_m) \\
&\quad + \xi_m^\rho\partial_\rho([A_\mu, W_n]) - \xi_n^\rho\partial_\rho([A_\mu, W_m]) \\
&\quad + D_\mu([W_m, W_n]) - [(\partial_\mu\xi_m^\rho)A_\rho + \xi_m^\rho\partial_\rho A_\mu, W_n] + [(\partial_\mu\xi_m^\rho)A_\rho, W_n] \\
&\quad + [(\partial_\mu\xi_n^\rho)A_\rho + \xi_n^\rho\partial_\rho A_\mu, W_m] - [(\partial_\mu\xi_n^\rho)A_\rho, W_m] \\
&= (\partial_\mu\xi_m^\rho)(\partial_\rho W_n) + \xi_m^\rho(\partial_\mu\partial_\rho W_n) - (\partial_\mu\xi_n^\rho)(\partial_\rho W_m) - \xi_n^\rho(\partial_\mu\partial_\rho W_m) \\
&\quad + \xi_m^\rho\partial_\rho([A_\mu, W_n]) - \xi_n^\rho\partial_\rho([A_\mu, W_m]) + (\partial_\mu\xi_m^\rho)[A_\rho, W_n] - (\partial_\mu\xi_n^\rho)[A_\rho, W_m] \\
&\quad + D_\mu([W_m, W_n]) - [D_\mu W_m, W_n] + [D_\mu W_n, W_m]
\end{aligned} \tag{3.7.18}
$$

by comparing with the left-hand side, we can see that the only difference is

$$
+ D_\mu([W_m, W_n]) - [D_\mu W_m, W_n] + [D_\mu W_n, W_m], \tag{3.7.19}
$$

however, this term is zero,

$$
\begin{aligned}
&+ D_\mu([W_m, W_n]) - [D_\mu W_m, W_n] + [D_\mu W_n, W_m] \\
&= \partial_\mu([W_m, W_n]) - [\partial_\mu W_m, W_n] - [W_m, \partial_\mu W_n] \\
&\ [A_\mu, [W_m, W_n]] + [W_m, [W_n, A_\mu]] + [W_n, [A_\mu, W_m]] \\
&= 0,
\end{aligned}
\tag{3.7.20}
$$

because of the Jacobi identity. Thus, we achieved our aims.

3.7.4 Equation (3.3.17)

We can prove

$$
D_\nu \mathbf{L}_\xi A_\mu - D_\mu \mathbf{L}_\xi A_\nu = \mathbf{L}_\xi F_{\nu\mu}. \tag{3.7.21}
$$

as follows;

$$
\begin{aligned}
D_\nu \mathbf{L}_\xi A_\mu - D_\mu \mathbf{L}_\xi A_\nu =& (\partial_\nu \partial_\mu \xi^\rho) A_\rho + (\partial_\mu \xi^\rho) \partial_\nu A_\rho + (\partial_\nu \xi^\rho)(\partial_\rho A_\mu) + \xi^\rho (\partial_\nu \partial_\rho A_\mu) \\
& - (\partial_\mu \partial_\nu \xi^\rho) - (\partial_\nu \xi^\rho)(\partial_\mu A_\rho) - (\partial_\mu \xi^\rho)(\partial_\rho A_\nu) - \xi^\rho (\partial_\mu \partial_\rho A_\nu) \\
& + (\partial_\mu \xi^\rho)[A_\nu, A_\rho] + \xi^\rho [A_\nu, \partial_\rho A_\mu] - (\partial_\nu \xi^\rho)[A_\mu, A_\rho] - \xi^\rho [A_\mu, \partial_\rho A_\nu] \\
=& (\partial_\mu \xi^\rho)(\partial_\nu A_\rho - \partial_\rho A_\nu + [A_\nu, A_\mu]) + (\partial_\nu \xi^\rho)(\partial_\rho A_\mu - \partial_\mu A_\rho + [A_\rho, A_\mu]) \\
& + \xi^\rho \partial_\rho (\partial_\mu A_\nu - \partial_\nu A_\mu) + \xi^\rho [A_\nu, \partial_\rho A_\mu] + \xi^\rho [\partial_\rho A_\nu, A_\mu] \\
=& (\partial_\mu \xi^\rho) F_{\rho\mu} + (\partial_\nu \xi^\rho) F_{\nu\rho} + \xi^\rho \partial_\rho F_{\nu\mu} \\
=& \mathbf{L}_\xi F_{\nu\mu}.
\end{aligned}
\tag{3.7.22}
$$

3.7.5 Equation (3.3.18)

With $\Psi_m = \xi^\mu_m A_\mu - W_m$, we prove

$$
\xi^\mu_m F_{\mu\nu} = -D_\nu \Psi_m. \tag{3.7.23}
$$

By using the symmetry equation,

$$
(\partial_\mu \xi^\rho_m) A_\rho + \xi^\rho_m \partial_\rho A_\mu = D_\mu W_m, \tag{3.7.24}
$$

we prove as follows;

$$\begin{aligned}\xi_m^\mu F_{\mu\nu} =&\xi_m^\mu(\partial_\mu A_\nu - \partial_\nu A_\mu - [A_\mu, A_\nu])\\ =&D_\nu W_m - (\partial_\nu \xi_m^\mu)A_\mu - \xi_m^\mu \partial_\nu A_\mu - [\xi_m^\mu A_\mu, A_\nu]\\ =&D_\nu W_m - \partial_\nu(\xi_m^\mu A_\mu) + [A_\nu, \xi_m^\mu A_\mu]\\ =&D_\nu(W_m - \xi_m^\mu A_\mu)\\ =&- D_\nu \Psi_m. \end{aligned}\tag{3.7.25}$$

3.7.6 Equation (3.3.20)

With $\Psi_m = \xi_m^\mu A_\mu - W_m$, we prove

$$\xi_m^\mu \xi_n^\nu F_{\mu\nu} = f_{mnp}\Psi_p - [\Psi_m, \Psi_n]. \tag{3.7.26}$$

To do this, first, we prove

$$\mathbf{L}_{\xi_m}\Psi_n - [W_m, \Psi_n] = f_{mnp}\Psi_p \tag{3.7.27}$$

Using the consistency condition;

$$\mathbf{L}_{\xi_m} W_n - \mathbf{L}_{\xi_n} W_m - [W_m, W_n] - f_{mnp} W_p = 0, \tag{3.7.28}$$

we get

$$\begin{aligned}\mathbf{L}_{\xi_m}\Psi_n - [W_m, \Psi_n] =&\mathbf{L}_{\xi_m}(\xi_n^\mu A_\mu) - \mathbf{L}_{\xi_m} W_n + [W_m, \xi_n^\mu A_\mu] + [W_m, W_n]\\ =&\xi_m^\rho(\partial_\rho \xi_n^\mu)A_\mu + \xi_m^\rho \xi_n^\mu \partial_\rho A_\mu + [W_m, \xi_n^\mu A_\mu] - \xi_n^\rho(\partial_\rho W_m) - f_{mnp} W_p. \end{aligned}\tag{3.7.29}$$

Moreover, using the symmetry equation,

$$(\partial_\mu \xi_m^\rho)A_\rho + \xi_m^\rho \partial_\rho A_\mu = \partial_\mu W_m + [A_\mu, W_m] \tag{3.7.30}$$

we continue this calculation to obtain

$$\begin{aligned}\mathbf{L}_{\xi_m}\Psi_n - [W_m, \Psi_n] =& \xi_m^\rho(\partial_\rho \xi_n^\mu)A_\mu + \xi_m^\rho \xi_n^\mu \partial_\rho A_\mu + [W_m, \xi_n^\mu A_\mu] + \xi_n^\rho[A_\rho, W_m]\\ &- f_{mnp} W_p - \xi_n^\rho(\partial_\rho \xi_m^\mu)A_\mu - \xi_n^\rho \xi_m^\mu(\partial_\mu A_\rho)\\ =& (\xi_m^\rho \partial_\rho \xi_n^\mu - \xi_n^\rho \partial_\rho \xi_m^\mu)A_\mu - f_{mnp} W_p = f_{mnp}\xi_p^\mu A_\mu - f_{mnp} W_p\\ =& f_{mnp}\Psi_p, \end{aligned}\tag{3.7.31}$$

so we get

$$\mathbf{L}_{\xi_m}\Psi_n - [W_m, \Psi_n] = f_{mnp}\Psi_p \tag{3.7.32}$$

Using this relation,

$$\xi^\mu_m F_{\mu\nu} = -D_\nu \Psi_m \tag{3.7.33}$$

$$\Psi_m = \xi^\mu_m A_\mu - W_m, \tag{3.7.34}$$

and the fact that the Lie derivative of scalar fields become

$$\mathbf{L}_{\xi_n}\Psi_m = \xi^\nu_n \partial_\nu \Psi_m, \tag{3.7.35}$$

we can prove (3.3.20) as follows;

$$\begin{aligned}\xi^\mu_m \xi^\nu_n F_{\mu\nu} =& -\xi^\nu_n(\partial_\nu\Psi_m + [A_\nu, \Psi_m]) = -\xi^\nu_n\partial_\nu\Psi_m - [\xi^\nu_n A_\nu, \Psi_m]\\ =& -\mathbf{L}_{\xi_n}\Psi_m + [W_n + \Psi_n, \Psi_m] = -[W_n, \Psi_m] - f_{nmp}\Psi_p + [W_n + \Psi_n, \Psi_m]\\ =& f_{mnp}\Psi_p - [\Psi_m, \Psi_n].\end{aligned} \tag{3.7.36}$$

3.7.7 Equation (3.3.27)

Rewriting W_m in

$$\mathbf{L}_{\xi_m} W_n(y^{\hat\alpha}) - \mathbf{L}_{\xi_n} W_m(y^{\hat\alpha}) + [W_m(y^{\hat\alpha}), W_n(y^{\hat\alpha})] - f_{mnp}W_p(y^{\hat\alpha}) = 0 \tag{3.7.37}$$

as

$$W_m = \xi^{\hat\alpha}_m W_{\hat\alpha}, \tag{3.7.38}$$

we prove

$$\partial_{\hat\alpha} W_{\hat\beta} - \partial_{\hat\beta} W_{\hat\alpha} + [W_{\hat\alpha}, W_{\hat\beta}] = 0. \tag{3.7.39}$$

We substitute $W_m = \xi^{\hat\alpha}_m W_{\hat\alpha}$ into above equation, we obtain

$$\begin{aligned}&\xi^{\hat\alpha}_m\partial_{\hat\alpha}(\xi^{\hat\beta}_n W_{\hat\beta}) - \xi^{\hat\alpha}_n\partial_{\hat\alpha}(\xi^{\hat\beta}_m W_{\hat\beta}) - [\xi^{\hat\alpha}_m W_{\hat\alpha}, \xi^{\hat\beta}_n W_{\hat\beta}] - f_{mnp}\xi^{\hat\beta} W_{\hat\beta}\\ &= \xi^{\hat\alpha}_m(\partial_{\hat\alpha}\xi^{\hat\beta}_n)W_{\hat\beta} - \xi^{\hat\alpha}_n(\partial_{\hat\alpha}\xi^{\hat\beta}_m)W_{\hat\beta} + \xi^{\hat\alpha}_m\xi^{\hat\beta}_n\partial_{\hat\alpha}W_{\hat\beta} - \xi^{\hat\alpha}_n\xi^{\hat\beta}_m\partial_{\hat\alpha}W_{\hat\beta} - \xi^{\hat\alpha}_m\xi^{\hat\beta}_n[W_{\hat\alpha}, W_{\hat\beta}] - f_{mnp}\xi^{\hat\beta}W_{\hat\beta}\end{aligned} \tag{3.7.40}$$

Here, noting

$$\xi_m^{\hat{\alpha}}(\partial_{\hat{\alpha}}\xi_n^{\hat{\beta}})W_{\hat{\beta}} - \xi_n^{\hat{\alpha}}(\partial_{\hat{\alpha}}\xi_m^{\hat{\beta}})W_{\hat{\beta}} - f_{mnp}\xi^{\hat{\beta}}W_{\hat{\beta}}$$
$$= ([\xi_m, \xi_n]^{\hat{\beta}} - f_{mnp}\xi^{\hat{\beta}})W_{\hat{\beta}} = 0, \quad (3.7.41)$$

we can show

$$\xi_m^{\hat{\alpha}}\xi_n^{\hat{\beta}}\partial_{\hat{\alpha}}W_{\hat{\beta}} - \xi_n^{\hat{\alpha}}\xi_m^{\hat{\beta}}\partial_{\hat{\alpha}}W_{\hat{\beta}} - \xi_m^{\hat{\alpha}}\xi_n^{\hat{\beta}}[W_{\hat{\alpha}}, W_{\hat{\beta}}] = 0$$
$$\to \xi_m^{\hat{\alpha}}\xi_n^{\hat{\beta}}(\partial_{\hat{\alpha}}W_{\hat{\beta}} - \partial_{\hat{\beta}}W_{\hat{\alpha}} + [W_{\hat{\alpha}}, W_{\hat{\beta}}]) = 0 \quad (3.7.42)$$

so, we can prove (3.3.27).

3.7.8 Equation (3.3.54)

$$\xi_m^{\hat{\alpha}}F_{\hat{\alpha}i} = -D_i\Psi_m = -D_i(W_m - \xi_m^{\hat{\alpha}}A_{\hat{\alpha}}) \quad (3.7.43)$$

Now, noting $W_m = 0$, by substituting the solution $A_{\hat{\alpha}}$ of the symmetry equation into above equation and multiplying both sides by $\xi_{m\hat{\beta}}$, we can calculate

$$\xi_{m\hat{\beta}}\xi_m^{\hat{\alpha}}F_{\hat{\alpha}i} = \xi_{m\hat{\beta}}\xi^{\hat{\alpha}}\tilde{\xi}_{n\hat{\alpha}}(-\partial_i\Phi_n - [A_i, \Phi_n])$$
$$\to F_{\hat{\beta}i} = \tilde{\xi}_{n\hat{\beta}}(-\partial_i\Phi_n - [A_i, \Phi_n]) \quad (3.7.44)$$

using $\xi_m^{\hat{\alpha}}\xi_{m\hat{\beta}} = \delta_{\hat{\beta}}^{\hat{\alpha}}$.

To satisfy constraint $F_{\omega i} = 0$, consider the case $\hat{\beta} = \omega$ for $F_{\hat{\beta}i} = \tilde{\xi}_{n\hat{\beta}}(-\partial_i\Phi_n - [A_i, \Phi_n]) = 0$ in the above equation. For $N \le N'$, we have $\tilde{\xi}_{n\omega} = 0$, so this is trivially satisfied, but for $m > N'$, we have $\tilde{\xi}_{n\omega} \neq 0$, so

$$\partial_i\Phi_n + [A_i, \Phi_n]) = 0 \quad (3.7.45)$$

is required.

3.7.9 Equation (3.3.55)

Multiplying both sides of

$$\xi_m^{\hat{\alpha}}\xi_n^{\hat{\beta}}F_{\hat{\alpha}\hat{\beta}} = f_{mnp}\Psi_p + [\Psi_m, \Psi_n] = f_{mnp}\xi_p^{\hat{\alpha}}\Phi_k\tilde{\xi}_{k\hat{\alpha}} + [\xi_m^{\hat{\alpha}}\Phi_k\tilde{\xi}_{k\hat{\alpha}}, \xi_n^{\hat{\beta}}\Phi_l\tilde{\xi}_{l\hat{\beta}}] \quad (3.7.46)$$

by $\xi_{m\hat{\gamma}}\xi_{n\hat{\delta}} = h_{\hat{\gamma}\hat{\rho}}h_{\hat{\delta}\hat{\sigma}}\xi_m^{\hat{\rho}}\xi_n^{\hat{\sigma}}$ yields

$$\xi_{m\hat{\gamma}}\xi_{n\hat{\delta}}\xi_m^{\hat{\alpha}}\xi_n^{\hat{\beta}}F_{\hat{\alpha}\hat{\beta}} = h_{\hat{\gamma}\hat{\rho}}h_{\hat{\delta}\hat{\sigma}}f_{mnp}\xi_m^{\hat{\rho}}\xi_n^{\hat{\sigma}}\xi_p^{\hat{\alpha}}\Phi_k\tilde{\xi}_{k\hat{\alpha}} + \xi_{m\hat{\gamma}}\xi_{n\hat{\delta}}\xi_m^{\hat{\alpha}}\xi_n^{\hat{\beta}}\tilde{\xi}_{k\hat{\alpha}}\tilde{\xi}_{l\hat{\beta}}[\Phi_k, \Phi_l] \tag{3.7.47}$$

Using $f_{mnp}\tilde{\xi}_m^{\hat{\alpha}}\tilde{\xi}_n^{\hat{\beta}}\tilde{\xi}_p^{\hat{\gamma}} = -f_{mnp}\xi_m^{\hat{\alpha}}\xi_n^{\hat{\beta}}\xi_p^{\hat{\gamma}}$, we get

$$\begin{aligned} F_{\hat{\gamma}\hat{\delta}} &= -f_{mnp}\tilde{\xi}_{m\hat{\gamma}}\tilde{\xi}_{n\hat{\delta}}\Phi_p + \tilde{\xi}_{m\hat{\gamma}}\tilde{\xi}_{n\hat{\delta}}[\Phi_m, \Phi_n] \\ &= -\tilde{\xi}_{m\hat{\gamma}}\tilde{\xi}_{n\hat{\delta}}(f_{mnp} - [\Phi_m, \Phi_n]). \end{aligned} \tag{3.7.48}$$

To satisfy constraint $F_{\tau\omega} = 0$ (ω, τ are indices of the subgroup R),

$$F_{\tau\omega} = -\tilde{\xi}_{m\tau}\tilde{\xi}_{n\omega}(f_{mnp} - [\Phi_m, \Phi_n]) = 0 \tag{3.7.49}$$

is required and to satisfy constraint $F_{\alpha\omega} = 0$,

$$F_{\alpha\omega} = -\tilde{\xi}_{m\alpha}\tilde{\xi}_{n\omega}(f_{mnp} - [\Phi_m, \Phi_n]) = 0 \tag{3.7.50}$$

is required. Thus,

$$\tilde{\xi}_{n\omega}(f_{mnp} - [\Phi_m, \Phi_n]) = 0 \tag{3.7.51}$$

is required for both $F_{\tau\omega}$ and $F_{\alpha\omega} = 0$ to be satisfied.

For $m \leq N'$, this equation is trivially satisfied because of $\tilde{\xi}_{n\omega} = 0$, but for $m > N'$,

$$f_{mnp} - [\Phi_m, \Phi_n] = 0 \tag{3.7.52}$$

is required because $\tilde{\xi}_{n\omega} \neq 0$. Note that m, p can take arbitrary values.

3.7.10 Equation (3.3.63)

On the subgroup R, with $r(y^\omega)$ the element of the group ($s(y^{\hat{\alpha}}) = r(y^\omega)s_0(y^\alpha)$ the element of the group S) and its generator as $\tilde{\xi}_m^\omega$, we have

$$\begin{aligned} &\mathbf{L}_{\tilde{\xi}_m} s = -J^m s = -J^m r s_0 \\ &= \tilde{\xi}_m^\omega(\partial_\omega r)s_0 \\ &\to -J^m r = \tilde{\xi}_m^\omega(\partial_\omega r), \end{aligned} \tag{3.7.53}$$

in the same way as it holds on S.

Multiplying both sides by $\tilde{\xi}_{m\tau}$ leads to

$$
\begin{aligned}
&-\tilde{\xi}_{m\tau}J^m r = \tilde{\xi}_{m\tau}\tilde{\xi}^{\omega}_m(\partial_\omega r) = \partial_\tau r \\
&\rightarrow (\partial_\omega r)r^{-1} = -\tilde{\xi}_{m\omega}J^m.
\end{aligned}
\tag{3.7.54}
$$

3.7.11 The Calculation of Sect. 3.3.2

We show the some formula.

$$
t_1 = \frac{i}{2}(\chi_\gamma + \chi_{-\gamma}),\quad t_2 = \frac{1}{2}(\chi_\gamma, \chi_{-\gamma}),\quad t_3 = \frac{i}{2}h_\gamma,\quad y = \frac{i}{2}h \tag{3.7.55}
$$

$$
h_\omega = \frac{2\omega_i}{\omega\cdot\omega}H_i,\quad h = \frac{2}{\sqrt{\gamma\cdot\gamma}}\frac{1}{\sqrt{\gamma_1^2+\gamma_2^2}}(\gamma_2 H_1 - \gamma_1 H_2) \tag{3.7.56}
$$

$$
[H_i, \chi_\omega] = \omega_i\chi_\omega,\quad [\chi_\omega, \chi_{-\omega}] = \frac{2\omega_i}{\omega\cdot\omega}H_i,\quad [\chi_\omega, \chi_\tau] = c_{\omega,\tau}\chi_{\omega+\tau} \tag{3.7.57}
$$

$$
\rightarrow [\chi_\gamma, \chi_\beta] = \pm\chi_\alpha,\quad [\chi_{-\gamma}, \chi_{-\beta}] = \mp\chi_{-\alpha},\quad [\chi_{-\gamma}, \chi_\alpha] = \pm\chi_\beta,\quad [\chi_\gamma, \chi_{-\alpha}] = \mp\chi_{-\beta} \tag{3.7.58}
$$

The $\alpha+\gamma$ and $\beta-\gamma$ are not roots.

$$
[t_1, \chi_\alpha] = \frac{i}{2}[\chi_{-\gamma}, \chi_\alpha] = +\frac{i}{2}\chi_\beta \tag{3.7.59}
$$

$$
[t_1, \chi_\beta] = \frac{i}{2}[\chi_\gamma, \chi_\beta] = +\frac{i}{2}\chi_\alpha \tag{3.7.60}
$$

$$
[t_2, \chi_\alpha] = -\frac{1}{2}[\chi_{-\gamma}, \chi_\alpha] = -\frac{1}{2}\chi_\beta = +\frac{i}{2}(+i\chi_\beta) \tag{3.7.61}
$$

$$
[t_2, \chi_\beta] = \frac{1}{2}[\chi_\gamma, \chi_\beta] = \frac{1}{2}\chi_\alpha = +\frac{i}{2}(-i\chi_\alpha) \tag{3.7.62}
$$

$$
[t_3, \chi_\alpha] = \frac{i}{2}\frac{2\gamma_i}{\gamma\cdot\gamma}[H_i, \chi_\alpha] = \frac{i}{2}\frac{2\gamma\cdot\alpha}{\gamma\cdot\gamma}\chi_\alpha = \frac{i}{2}\langle\alpha, \gamma\rangle\chi_\alpha = \frac{i}{2}\chi_\alpha \tag{3.7.63}
$$

$$
[t_3, \chi_\beta] = \frac{i}{2}\langle\beta, \gamma\rangle\chi_\beta = \frac{i}{2}\langle\alpha-\gamma, \gamma\rangle\chi_\beta = \frac{i}{2}(1-2)\chi_\beta = -\frac{i}{2}\chi_\beta \tag{3.7.64}
$$

$$
[y, \chi_\alpha] = \frac{i}{2}h_i[H_i, \chi_\alpha] = \frac{i}{2}h_i\alpha_i\chi_\alpha = \frac{i}{2}\tan\theta\chi_\alpha,\quad (h_i\alpha_i = \tan\theta) \tag{3.7.65}
$$

$$
[y, \chi_\beta] = \frac{i}{2}h_i[H_i, \chi_\beta] = \frac{i}{2}h_i\beta_i\chi_\beta = \frac{i}{2}h_i(\alpha_i - \gamma_i)\chi_\beta = \frac{i}{2}\tan\theta\chi_\beta \tag{3.7.66}
$$

Here, we use

$$(h_\alpha, h_\alpha) = \frac{2\alpha_i}{\alpha\cdot\alpha}\frac{2\alpha_j}{\alpha\cdot\alpha}(H_i, H_j) = \frac{4}{\alpha\cdot\alpha} \quad (3.7.67)$$

$$(h_\alpha, h_\beta) = \frac{2\alpha_i}{\alpha\cdot\alpha}\frac{2(\alpha_j - \gamma_j)}{\beta\cdot\beta}\delta_{ij} = \frac{4}{\alpha\cdot\alpha} - \frac{2}{\alpha\cdot\alpha}\langle\gamma, \alpha\rangle = \frac{2}{\alpha\cdot\alpha} \quad (3.7.68)$$

$$(h_\alpha, \chi_\gamma) = (h_\beta, \chi_\gamma) = \frac{2\alpha_i}{\alpha\cdot\alpha}(H_i, \chi_\gamma) = 0 \quad (3.7.69)$$

to above calculate.

References

1. Hiroyuki H, Masaki M (2008) Baryons and the Chern-Simons term in holographic QCD with three flavors. Prog Theor Phys 119:461–490
2. Pak HCL, Shigeki S (2017) Chern-Simons five-form and holographic baryons. Phys Rev D 95(12):126007
3. Daisuke F, Atsushi H (2020) Heavy baryons in holographic QCD with higher dimensional degrees of freedom. Phys Rev D 101(12):126008
4. Callan CG Jr, Klebanov IR (1985) Bound State Approach to Strangeness in the Skyrme Model. Nucl Phys B 262:365–382
5. Callan CG Jr, Hornbostel K, Klebanov IR (1988) Baryon masses in the bound state approach to strangeness in the skyrme model. Phys Lett B 202:269–275
6. Takashi E, Atsushi H (2017) Kaon-Nucleon scattering states and potentials in the Skyrme model. Phys Rev D 96(5):054002
7. Takashi E, Atsushi H (2016) Kaon-Nucleon systems and their interactions in the Skyrme model. Phys Rev D 94(3):034022
8. 江添貴之. Skyrme 模型の束縛法を用いた K 中間子核子間相互作用の導出, 大阪大学理学研究科修士論文
9. Yizhuang L, Ismail Z (2017) Heavy and strange holographic baryons. Phys Rev D 96(5):056027
10. Yizhuang L (2017) Heavy baryons and their exotics from instantons in holographic qcd. Phys Rev D 95(11) (2017)
11. Yizhuang L (2017) Heavy and strange holographic baryons. Phys Rev D 96(5)
12. Yizhuang L, Ismail Z (2017) Heavy baryons and their exotics from instantons in holographic QCD. Phys Rev D 95(11):116012
13. Yizhuang L, Nowak MA, Zahed I (2021) Holographic charm and bottom pentaquarks. I. Mass spectra with spin effects. Phys Rev D 104(11):114021 (2021)
14. Takaaki I (2011) Toward bound-state approach to strangeness in holographic QCD. Phys Lett B 695:392–396
15. Tadakatsu S, Shigeki S (2005) Low energy hadron physics in holographic QCD. Prog Theor Phys 113:843–882
16. Tadakatsu S, Shigeki S (2005) More on a holographic dual of QCD. Prog Theor Phys 114:1083–1118
17. Koji H, Takayuki H, Feng-Li L, Ho-Ung Y (2008) Quark mass deformation of holographic massless QCD. JHEP 07:089
18. Koji H, Norihiro I, Takaaki I, Daisuke K (2010) Three-flavor quark mass dependence of baryon spectra in holographic QCD. Phys Lett B 691:65–71
19. Koji H (2010) Hadron spectra and quark mass dependence in holographic QCD. Prog Theor Phys Suppl 186:211–215

20. 村田 仁樹. Hadrons in holographic QCD, 京都大学大学院理学研究科 修士論文
21. Yabu H, Ando K (1988) A New Approach to the SU(3) Skyrme Model. Nucl Phys B 301:601
22. Edward W (1983) Global aspects of current algebra. Nucl Phys B 223:422–432
23. Edward W (1983) Current algebra, baryons, and quark confinement. Nucl Phys B 223:433–444
24. Guadagnini E (1984) Baryons as solitons and mass formulae. Nucl Phys B 236:35–47
25. Sanjay J, Wadia Spenta R (1985) Large-N baryons: collective coordinates of the topological soliton in the SU(3) chiral model. Nucl Phys B 258:713–725
26. Mazur Pawel O, Nowak Maciej A, Michal Praszalowicz (1984) SU(3) extension of the skyrme model. Phys Lett B 147:137–140
27. Chemtob M (1985) Skyrme model of baryon octet and decuplet. Nucl Phys B 256:600–608
28. McNamee PSJ, Chilton F (1964) Tables of Clebsch-Gordon coefficients of SU3. Rev Mod Phys 36:1005–1024
29. Ofer A, David K (2008) Holographic duals of long open strings. Phys Rev D 78:026005
30. Koji H, Takayuki H, Akitsugu M (2007) Holographic QCD and pion mass. JHEP 06:020
31. Roberto C, Elias K, Angel P (2007) Chiral symmetry breaking as open string tachyon condensation. Nucl Phys B 787:98–134
32. Avinash D, Partha N (2008) Sakai-Sugimoto model, tachyon condensation and chiral symmetry breaking. JHEP 01:055
33. Avinash D, Partha N (2008) Tachyon condensation and quark mass in modified Sakai-Sugimoto model. Phys Rev D 78:066021
34. Forgacs P, Manton NS (1980) Space-time symmetries in gauge theories. Commun Math Phys 72:15
35. Manton NS (1979) A new six-dimensional approach to the Weinberg-Salam model. Nucl Phys B 158:141–153
36. Schwarz AS, Tyupkin YS (1981) Dimensional reduction of the gauge field theory. Nucl Phys B 187:321–332
37. Luciani JF (1978) Space-time geometry and symmetry breaking. Nucl Phys B 135:111
38. Manton NS, Sutcliffe P (2004) Topological solitons. Cambridge Monographs on Mathematical Physics. Cambridge University Press
39. Pons JM, Russo JG, Talavera P (2004) Semiclassical string spectrum in a string model dual to large N QCD. Nucl Phys B 700:71–88
40. Hiroyuki H, Tadakatsu S, Shigeki S, Shinichiro Y (2007) Baryons from instantons in holographic QCD. Prog Theor Phys 117:1157
41. Atiyah MF, Manton NS (1989) Skyrmions from instantons. Phys Lett B 222:438–442
42. Manohar AV, Wise MB (2000) Heavy quark physics, vol 10
43. Dominguez CA, Paver N (1987) Leptonic decay constants of charm and beauty mesons in QCD. Phys Lett B 197:423. [Erratum: Phys.Lett.B 199, 596 (1987)]
44. Adkins Gregory S, Nappi Chiara R, Edward W (1983) Static properties of nucleons in the skyrme model. Nucl Phys B 228:552
45. Takayama M, Toki H, Hosaka A (1999) Systematics of the SU(3) baryon spectra and deformed oscillator quark model. Prog Theor Phys 101:1271–1283
46. Arifi AJ, Nagahiro H, Hosaka A, Tanida K (2020) Roper-like resonances with various flavor contents and their two-pion emission decays. Phys Rev D 101(11):111502
47. Roel A et al (2015) Observation of $J/\psi p$ Resonances Consistent with Pentaquark States in $\Lambda_b^0 \to J/\psi K^- p$ Decays. Phys Rev Lett 115:072001
48. Roel A et al (2019) Observation of a narrow pentaquark state, $P_c(4312)^+$, and of two-peak structure of the $P_c(4450)^+$. Phys Rev Lett 122(22):222001
49. Nakahara M (2003) Geometry, topology and physics
50. Nair VP (2005) Quantum field theory: a modern perspective
51. Wess J, Zumino B (1971) Consequences of anomalous Ward identities. Phys Lett B 37:95–97

Chapter 4
Properties of Nucleon Resonances

Abstract We show how to calculate the electromagnetic transition amplitude and the decay width of one pion emission of nucleon resonance using the Sakai-Sugimoto model, and how to define the electromagnetic and axial currents using the GKP-Witten relation. An alternative definition of the current based on the Skyrmion derived from this model is also presented. The electromagnetic transition amplitudes and decay widths obtained from the two currents are compared with experimental data.

In this chapter, we use the Sakai-Sugimoto model to calculate the electromagnetic transition amplitudes and decay widths of several low-lying nucleon excited states. These observables strongly reflect the internal structure of hadrons, leading to a further understanding of the properties of hadron resonances and a deeper understanding of low-energy QCD. In particular, since the Sakai-Sugimoto model represents hadron resonances by utilizing the geometry of extra dimensions, it is useful to investigate the dynamical properties in order to verify the validity of this picture.

First, in the next section, we define the chiral current. In the Sakai-Sugimoto model, the current has been defined from two points of view. Therefore, we discuss these points and review two definitions. After that, in Sect. 4.2, we will explain the analysis of static properties of nucleons, and finally, in Sect. 4.3, we will introduce our recent works on the analysis of dynamical properties of nucleon resonances.

4.1 The Definition of the Chiral Current in the Sakai-Sugimoto Model

In order to investigate the dynamical properties of nucleon resonances, it is necessary to define the chiral current. In the Sakai-Sugimoto model, the chiral current is defined using the GKP-Witten relation. However, because this method defines the chiral current in the boundary of the bulk theory, the matrix element of the current by baryon states vanishes in the case where there is a BPST instanton solution (baryon) with SO(4) symmetry in a four-dimensional space. This is attributed to the fact that by procesure of the $1/\lambda$ expansion, we ignore the effects of the warp factors $h(z)$ and $k(z)$, and obtain the classical solution. Therefore, taking into account the effect of the

D. Fujii, *Dynamical Properties of Baryon Resonances in the Holographic QCD*, Springer Theses, https://doi.org/10.1007/978-981-97-0755-3_4

warp factor, we determine the asymptotic behavior of the soliton solution to allow the calculation of well-defined currents. Alternatively, by considering the bulk theory as a five-dimensional hadron effective theory, we can have another definition of currents. The former is a proper method from the viewpoint of AdS/CFT correspondence, which has been confirmed to explain the nucleon properties well [1]. On the other hand, the latter method, which also reproduce the nucleon properties well [2], has an unclear connection with the AdS/CFT correspondence. However, considering that the five-dimensional hadron model proposed in Ref. [3] leads to the same action as the Sakai-Sugimoto model, the bulk theory of the Sakai-Sugimoto model is expected to be seen as a hadron effective theory in a more extensive sense rather than being restricted to AdS/CFT, which ensures the validity of the latter current.

4.1.1 GKP-Witten Relation

In this subsection, we introduce a method for computing physical quantities related to QCD using holographic approach, called the GKP-Witten relation. The GKP-Witten relation regards the field in the bulk as external fields of the gauge theory existing at its boundary and allows the analysis of physical quantities coupled to it [4, 5] (Ref. [6] as a very useful review.).

The AdS/CFT proposes the equivalence of gauge theory and string theory. This specifically means that each of the partition functions is equivalent,

$$Z_{\text{gauge}} = Z_{\text{string}}. \tag{4.1.1}$$

Furthermore, as explained in Sect. 2.1, in the large 't Hooft coupling and large N_c limit, string theory is justified to be treated as a classical gravity theory (in general, the matter fields are also coupled). Therefore, the formula (4.1.1) reduces to

$$Z_{\text{gauge}}^{\lambda, N_c \gg 1} = e^{i \underline{S}_{\text{gravity+matter}}} \tag{4.1.2}$$

Here, $\underline{S}_{\text{gravity}}$ is an on-shell action. Now, when we add the external source $\phi^{(\infty)}$, and the operator combined with it is $\mathcal{O}$, it reduces to

$$\left\langle \exp\left(i \int \phi^{(\infty)} \mathcal{O}\right)\right\rangle = e^{i \underline{S}_{\text{gravity+matter}}[\phi|_{u=\infty} = \phi^{(\infty)}]}. \tag{4.1.3}$$

This external source is not only a scalar field, but can also be a vector or tensor for Lorenz indexes. Also, $\phi(x^\mu, u)$ is a field in gravity+matter theory, and $\phi^{(\infty)}$ means the value at the AdS boundary $u \to \infty$, which is explained below. Using this equivalence of pertition functions, we obtain the GKP-Witten relation;

$$\langle \mathcal{O} \rangle = \frac{\delta \underline{S}_{\text{gravity+matter}}[\phi|_{u=\infty} = \phi^{(\infty)}]}{\delta \phi^{\infty}}. \tag{4.1.4}$$

It is possible to have as many different gravity theories as there are corresponding gauge theories. In this section, we use the equivalence between gauge theory on the D3-brane in the flat spacetime and a black 3-brane, introduced in Sect. 2.5, to calculate physical quantities on the gauge theory side. A calculation almost similar to this simple example is also performed when calculating the chiral current, which is the main subject of this chapter.

4.1.1.1 For Example: Dilaton Field

■ Gauge theory side

We consider a gauge theory on D3-brane under flat spacetime. By considering perturbations of the dilaton field and substituting $S_{\rm DBI}$ action (2.5.1) for $e^{-\phi} = e^{-\phi_0 + \tilde{\phi}^{(\infty)}}$, we obtain

$$\begin{aligned} S_{\rm DBI} &= -T_p \int d^{3+1}\xi e^{-\phi_0+\tilde{\phi}^{(\infty)}} \sqrt{-\det(P[g]_{ab} + 2\pi\alpha' F_{ab})} \\ &= -\frac{T_p}{g_s} \int dt d^3 x e^{-\tilde{\phi}^{(\infty)}} \sqrt{-\det(P[g]_{ab} + 2\pi\alpha' F_{ab})} \\ &\sim -\frac{1}{4g_{YM}^2} \int dt d^3 x \mathrm{Tr}(F^2) (= S_{\rm YM}) + \frac{1}{4g_{YM}^2} \int dt d^3 x \tilde{\phi}^{(\infty)} \mathrm{Tr}(F^2). \end{aligned} \tag{4.1.5}$$

Here, there is also a contribution from the supersymmetric partner in this theory, but for the simplicity of the discussion, we omit it in the following. Thus, from the gauge theory side, we obtain the pertition function

$$Z_{\rm gauge} = \left\langle \exp\left[i\left(S_{\rm YM} + \frac{1}{4g_{YM}^2} \int dt d^3 x \tilde{\phi}^{(\infty)} \mathrm{Tr}(F^2) \right)\right]\right\rangle \tag{4.1.6}$$

in the presence of the external field $\tilde{\phi}$.

■ Gravity theory side

Considering the perturbation $\phi = \phi_0 + \tilde{\phi}$, $\tilde{\phi}$ is an infinitesimal quantity, so if we take only its leading term from (2.5.6), the equation of motion it follows is the Klein-Gordon equation,

$$\frac{1}{\sqrt{-g}} \partial_\mu (\sqrt{-g} g^{\mu\nu} \partial_\nu \tilde{\phi}) = 0 \tag{4.1.7}$$

Now, for simplicity, if we consider only the u-dependence and concretely substitute the metric (2.5.21), we obtain

$$\partial_u (u^5 g \partial_u \tilde{\phi}(u)) = 0. \tag{4.1.8}$$

It is easy to find this solution, which is written as

$$\tilde{\phi}(u) = \phi^{(0)} + \phi^{(1)} u^{-4}, \tag{4.1.9}$$

using the two integral constants $\phi^{(0)}$, $\phi^{(1)}$. Here, the second term falls faster than the first term at the boundary of the AdS spacetime at $u \to \infty$, which is called the normalizable mode. On the other hand, the first term is called the non-normalizable mode. This integral constant will be determined later.

Substituting the obtained classical solution into action (2.5.6), we calculate the on-shell action $\underline{S}_{\text{gravity+matter}}$. The on-shell action of $\tilde{\phi}$ is

$$\begin{aligned}
\underline{S}_{\text{gravity+matter}} &= -\frac{1}{2}\int dtd^3xdu\sqrt{-g}g^{\mu\nu}\partial_\mu\tilde{\phi}\partial_\nu\tilde{\phi} \\
&= \frac{1}{2}\int dtd^3xdu\tilde{\phi}\partial_u(u^5\partial_u\tilde{\phi}) - \frac{1}{2}\int dtd^3xu^5\tilde{\phi}\partial_u\tilde{\phi}\Big|_{u=0}^{u=\infty} \\
&= \int dtd^3x2\phi^{(0)}\phi^{(1)}\Big|_{u=\infty} - \int dtd^3x(\phi^{(0)}\phi^{(1)} + \phi^{(1)2}u^{-4})\Big|_{u=0}.
\end{aligned} \tag{4.1.10}$$

Here, by using the equation of motion from the second to the third line, we see that only the surface term remains. In the strongly coupled, large N_c gauge theory is often said to live on the boundary of gravity theory in this sense.

The two integral constants are determined as follows. First, it is understood that for the on-shell action to be finite, we only allow the solution $\phi^{(1)} = 0$ so that the contribution from the interior $u = 0$ of the bulk does not diverge. Next, solving the equation of motion of this dilaton field in the bulk under the boundary condition $\tilde{\phi}|_{u=\infty} = \tilde{\phi}^\infty$, we can determine that $\phi^{(0)} = \tilde{\phi}^\infty$. By matching the dynamical field in the bulk (with respect to u) with the external field in the boundary where the gauge theory lives, the physical quantity on the gauge theory side can be obtained by the analysis of the classical gravity theory (solving the equations of motion satisfying the boundary conditions).

From the above, using (4.1.4), we can calculate

$$\left\langle \mathrm{Tr}(F^2)\right\rangle = \frac{\delta Z_{\text{gauge}}}{\delta\tilde{\phi}^\infty} = \frac{\delta \exp[-i\underline{S}_{\text{gravity+matter}}]}{\delta\tilde{\phi}^\infty} \propto \phi^{(1)} = 0. \tag{4.1.11}$$

This is the quantity corresponding to the gluon condensate, and from the supersymmetry of the $\mathcal{N} = 4$ supersymmetric YM gauge theory we are considering now, this result was necessarily obtained.

4.1.2 The Chiral Current from the GKP-Witten Relation

4.1.2.1 GKP-Witten Relation for Chiral Current

In this section, we calculate the currents $\mathcal{J}^{\mu}_{L/R}$ associated with the chiral symmetry $U(N_f)_L \times U(N_f)_R$.

■ QCD side

If we introduce an external gauge field $\mathcal{A}_{L/R\mu}$ associated with the chiral symmetry $U(N_f)_L \times U(N_f)_R$, their linear terms are combined with the chiral current, as in

$$S_{\mathrm{QCD}}|_{\mathcal{O}(\mathcal{A}_L,\mathcal{A}_R)} = -2\int d^4x\mathrm{tr}\big(\mathcal{A}_{L\mu}\mathcal{J}^{\mu}_L + \mathcal{A}_{R\mu}\mathcal{J}^{\mu}_R\big). \tag{4.1.12}$$

Thus, the chiral current is calculated as

$$\left\langle \mathcal{J}^{\mu}_{L/R}\right\rangle \propto \frac{\delta Z_{\mathrm{QCD}}}{\delta \mathcal{A}_{L/R\mu}}. \tag{4.1.13}$$

■ Gravity+matter theory side

On the gravity+matter theory side, it is necessary to solve the equation of motion under the boundary condition $\mathcal{A}_{\mu}(x^{\mu}, z \to \pm\infty) = \mathcal{A}_{R/L\mu}(x^{\mu})$. For this purpose, $SU(N_f)$ gauge field $\mathcal{A}_{\alpha}$ is decomposed into normalizable mode $\mathcal{A}^{cl}_{\alpha}$ and non-normalizable mode $\delta\mathcal{A}_{\alpha}$, as follow;

$$\mathcal{A}_{\alpha}\left(x^{\mu}, z\right) = \mathcal{A}^{cl}_{\alpha}\left(x^{\mu}, z\right) + \delta\mathcal{A}_{\alpha}\left(x^{\mu}, z\right). \tag{4.1.14}$$

The non-normalizable mode are related to the left and right gauge fields in the four dimensional space at $z \to \pm\infty$,

$$\begin{aligned}
\delta\mathcal{A}_{\mu}\left(x^{\nu}, z \to +\infty\right) &= \mathcal{A}_{L\mu}\left(x^{\nu}\right), \\
\delta\mathcal{A}_{\mu}\left(x^{\nu}, z \to -\infty\right) &= \mathcal{A}_{R\mu}\left(x^{\nu}\right).
\end{aligned}$$

The non-normalizable mode $\delta\mathcal{A}_{\alpha}$ will be obtained later. Substituting this field into the action, the coefficients of the first order in $\mathcal{A}_{L\mu}$, $\mathcal{A}_{R\mu}$ is identified with the left and right currents $\langle\mathcal{J}^{\mu}_L\rangle$, $\langle\mathcal{J}^{\mu}_R\rangle$ from on-shell action $\underline{S}_{\mathrm{bulk}}$ with the sign properly taken into account,

$$\begin{aligned}
\underline{S}_{\mathrm{bulk}}|_{\mathcal{O}(\mathcal{A}_L,\mathcal{A}_R)} &= \kappa\int d^4x\left[2\mathrm{tr}\left(\delta\mathcal{A}^{\mu}k\left(z\right)\mathcal{F}^{cl}_{\mu z}\right)\right]^{z=+\infty}_{z=-\infty}, \\
&= -2\int d^4x\mathrm{tr}\left(\mathcal{A}_{L\mu}\langle\mathcal{J}^{\mu}_L\rangle + \mathcal{A}_{R\mu}\langle\mathcal{J}^{\mu}_R\rangle\right).
\end{aligned} \tag{4.1.15}$$

where

$$\begin{aligned}\left\langle \mathcal{J}_L^{\mu} \right\rangle &= -\kappa \left(k\left(z\right) \mathcal{F}_{\mu z}^{cl} \right) \Big|_{z=+\infty}, \\ \left\langle \mathcal{J}_R^{\mu} \right\rangle &= +\kappa \left(k\left(z\right) \mathcal{F}_{\mu z}^{cl} \right) \Big|_{z=-\infty}.\end{aligned} \tag{4.1.16}$$

The vector and axial-vector currents are then obtained by

$$\begin{aligned}\left\langle \mathcal{J}_V^{\mu} \right\rangle &= \left\langle \mathcal{J}_L^{\mu} \right\rangle + \left\langle \mathcal{J}_R^{\mu} \right\rangle = -\kappa \left[k\left(z\right) \mathcal{F}_{\mu z}^{cl} \right]_{z=-\infty}^{z=+\infty} \\ \left\langle \mathcal{J}_A^{\mu} \right\rangle &= \left\langle \mathcal{J}_L^{\mu} \right\rangle - \left\langle \mathcal{J}_R^{\mu} \right\rangle = -\kappa \left[\psi_0\left(z\right) k\left(z\right) \mathcal{F}_{\mu z}^{cl} \right]_{z=-\infty}^{z=+\infty},\end{aligned} \tag{4.1.17}$$

with $\psi_0\left(z\right) = \left(2/\pi\right) \arctan z$.

The baryon number current is defined as

$$\left\langle J_B^{\mu} \right\rangle = \frac{2}{N_c} \langle \hat{J}_V^{\mu} \rangle = -\frac{2}{N_c} \kappa \Big[k(z) \hat{F}^{\mu z} \Big]_{z=-\infty}^{z=+\infty}. \tag{4.1.18}$$

Then, baryon number density is given by

$$\left\langle J_B^{0} \right\rangle = \frac{2}{N_c} \hat{J}_V^{\mu} = -\frac{2}{N_c} \kappa \Big[k(z) \hat{F}^{0z} \Big]_{z=-\infty}^{z=+\infty} = -\frac{2}{N_c} \kappa \int dz \partial_z \big(k(z) \hat{F}^{0z} \big). \tag{4.1.19}$$

Here, by using equation of motion, it becomes

$$\left\langle J_B^{0} \right\rangle = -\frac{1}{64\pi^2} \int dz \epsilon^{0MNPQ} F_{MN}^a F_{PQ}^a + \text{(total derivative)}. \tag{4.1.20}$$

Furthermore, integrating this in space, it coincides with the instanton number, as expected.

In the following, we will derive the four-dimensional effective action including the external field by mode expansion and see how $\left\langle \mathcal{J}_{V,\mu} \right\rangle$ and $\left\langle \mathcal{J}_{A,\mu} \right\rangle$ are written by the meson field, which is a component of this action. By using $\mathcal{V}_{V/A,\mu} = \mathcal{A}_{L,\mu} \pm \mathcal{A}_{R,\mu}$, this effective action is written as

$$S|_{\mathcal{O}(\mathcal{A}_L, \mathcal{A}_R)} = -2 \int d^4x \operatorname{tr} \big(\mathcal{V}_{V,\mu} \langle \mathcal{J}_V^{\mu} \rangle + \mathcal{V}_{A,\mu} \langle \mathcal{J}_A^{\mu} \rangle \big). \tag{4.1.21}$$

Now, as in (2.3.29), we perform the mode expansion of the gauge fields as

$$\begin{aligned}\mathcal{A}_{\mu}(x, z) &= \mathcal{A}_{L\mu}(x) \psi_+(z) + \mathcal{A}_{R\mu}(x) \psi_-(z) \\ &\quad + \sum_{n=1}^{\infty} v_{\mu}^n(x) \psi_{2n-1}(z) + \sum_{n=1}^{\infty} a_{\mu}^n(x) \psi_{2n}(z),\end{aligned} \tag{4.1.22}$$

$$\mathcal{A}_z(x, z) = \Pi(x) \phi_0(z), \tag{4.1.23}$$

$$\psi_{\pm}(z) = \frac{1}{2}(1 \pm \psi_0(z)), \quad \psi_0(z) = \frac{2}{\pi} \arctan z,$$

then substitute it into the action to get the 4-dimensional action

$$S|_{\mathcal{O}(\mathcal{A}_L,\mathcal{A}_R)} = \int d^4x 2\mathrm{tr}\Big(\mathcal{V}_{V,\mu}\sum_{n=1}^{\infty} g_{v^n} v_\mu^n + \mathcal{V}_{A,\mu}\Big(\sum_{n=1}^{\infty} g_{a^n} a_\mu^n + f_\pi \partial_\mu \Pi\Big)\Big), \tag{4.1.24}$$

where f_π is the pion decay constant and g_{v^n} and g_{a^n} the decay constant of the vector and axial vector meson, respectively, giving as

$$f_\pi = 2\sqrt{\frac{\kappa}{\pi}},\ g_{v^n} = -2\kappa(k(z)\partial_z\psi_{2n-1})|_{z=+\infty},\ g_{a^n} = -2\kappa(k(z)\partial_z\psi_{2n})|_{z=-\infty}. \tag{4.1.25}$$

It should be emphasized that in the Sakai-Sugimoto model, these constants are not parameters but quantities determined by the functions $\psi_n(z)$ of the complete system and the warp factor $h(z), k(z)$, which reflects the QCD information. For later use, the following expressions are also given;

$$g_{v^n} = \lambda_{2n-1}\kappa \int dz h(z)\psi_{2n-1},\ g_{a^n} = \lambda_{2n}\kappa \int dz h(z)\psi_{2n}\psi_0. \tag{4.1.26}$$

From this action, by reading $\mathcal{J}_{V,\mu}$ and $\mathcal{J}_{A,\mu}$, we obtain the following expression,

$$\langle \mathcal{J}_{V,\mu}\rangle = -\sum_{n=1}^{\infty} g_{v^n} v_\mu^n,\ \langle \mathcal{J}_{A,\mu}\rangle = -f_\pi\partial_\mu\Pi - \sum_{n=1}^{\infty} g_{a^n} a_\mu^n, \tag{4.1.27}$$

which show that vector currents are denoted by vector mesons only, indicating vector meson dominance.

4.1.2.2 The Asymptotic Behavior of the Instanton Solutions

The quantity we now desire to compute is the matrix element of the chiral current with baryons. We will accomplish this by finding a classical solution of the bulk theory such that the baryon number current is nontrivial, according to Ref. [1]. An easy way to immediately consider is to adopt the BPST instanton solution of the $SU(N_f)$ flavor gauge field in the bulk matter fields. However, the above current defined by $z \to \infty$ vanishes when the BPST instanton solution with SO(4) symmetry is substituted. This is because the effect of the warp factor in the z direction is neglected to solve the classical solutions at the large λ limit. Therefore, in the following, we will consider asymptotic behavior in $z \to \infty$ of the instanton solution.

We will determine the asymptotic behavior of the soliton solution, which allows to evaluate a well-defined of the $U(N_f)_L \times U(N_f)_R$ chiral symmetry. Without $1/\lambda$ expansion, the equations of motion are

$$2\kappa\left[\mathcal{D}_\nu(h(z)\mathcal{F}^{\nu\mu}) + \mathcal{D}_z(k(z)\mathcal{F}^{z\mu})\right] + \frac{N_c}{32\pi^2}\epsilon^{\mu NPQR}\mathcal{F}_{NP}\mathcal{F}_{QR} = 0, \quad (4.1.28)$$

$$2\kappa\mathcal{D}_\mu(k(z)\mathcal{F}^{\mu z}) + \frac{N_c}{32\pi^2}\epsilon^{\mu\nu\rho\sigma}\mathcal{F}_{\mu\nu}\mathcal{F}_{\rho\sigma} = 0, \quad (4.1.29)$$

with $N, P, Q, R = 0, 1, 2, 3, z$. Let us consider time-dependent gauge configuration in the moduli space. We have already discussed in Sect. 2.4.2, we can give a time dependence to the classical solutions of the gauge field as follows;

$$A_M = V A_M^{\rm cl} V^{-1} - iV\partial_M V^{-1}, \quad (4.1.30)$$

where V and W are related as

$$\Phi \equiv W^{-1}\Delta A_0 W - iW^{-1}\dot{W} = -iV^{-1}\dot{V}. \quad (4.1.31)$$

In the following transformation of the equation, this notation is more convenient, so we will use $V(t, \vec{x}, z)$. Here, it is useful to perform the following gauge transformation;

$$A_\alpha \to A_\alpha^G \equiv GA_\alpha G^{-1} - iG\partial_\alpha G^{-1}, \quad (4.1.32)$$

with $G = Wg^{-1}V^{-1}$. Then, the gauge fields are

$$A_0^G = -i(1 - f(\xi))W\dot{W}^{-1} + i(1 - f(\xi))\dot{X}^M W(g^{-1}\partial_M g)W^{-1}, \quad (4.1.33)$$

$$A_M^G = -i(1 - f(\xi))W(g^{-1}\partial_M g)W^{-1}, \quad (4.1.34)$$

where this choice of gauge is helpful in considering asymptotic behavior.

Now, we treat the $U(1)$ part as a perturbation under the background of the gauge configuration of the $SU(2)$ instanton with asymptotic behavior. The leading contribution to the $U(1)$ part is obtained by solving the following linearized equation of motion.

$$\partial_M\partial^M\hat{A}^0 = \frac{3}{\pi^2 a\lambda}\frac{\rho^4}{(\xi^2+\rho^2)^4} \quad (4.1.35)$$

$$\partial_M\partial^M\hat{A}^i = \frac{3}{\pi^2 a\lambda}\frac{\rho^4}{(\xi^2+\rho^2)^4}\left(\dot{X}^i + \frac{\chi^a}{2}(\epsilon^{iaj}x^j - \delta^{ia}z) + \frac{\dot{\rho}x^i}{\rho}\right) \quad (4.1.36)$$

$$\partial_M\partial^M\hat{A}_z = \frac{3}{\pi^2 a\lambda}\frac{\rho^4}{(\xi^2+\rho^2)^4}\left(\dot{Z} + \frac{\chi^a x^a}{2} + \frac{\dot{\rho}z}{\rho}\right) \quad (4.1.37)$$

Here we have substituted (4.1.33) and (4.1.34) into the equations of motion for (4.1.28) and (4.1.29) with the warp factors $k(z) = h(z) = 1$. We also neglected the terms involving ∂_0^2 because we are interested in slowly moving solitons. The regular solution is found as follows.

$$\hat{A}_0 = \frac{1}{8\pi^2 a\lambda}\frac{1}{\xi^2}\left[1 - \frac{\rho^4}{(\xi^2+\rho^2)^2}\right] = \frac{1}{8\pi^2 a\lambda}\frac{\xi^2+2\rho^2}{(\xi^2+\rho^2)^2} \tag{4.1.38}$$

$$\hat{A}_i = -\frac{1}{8\pi^2 a\lambda}\left[\frac{\xi^2+2\rho^2}{(\xi^2+\rho^2)^2}\dot{X}^i + \frac{\rho^2}{(\xi^2+\rho^2)^2}\left(\frac{\chi^a}{2}(\epsilon^{iaj}x^j - \delta^{ia}z) + \frac{\dot{\rho}x^i}{\rho}\right)\right] \tag{4.1.39}$$

$$\hat{A}_z = -\frac{1}{8\pi^2 a\lambda}\left[\frac{\xi^2+2\rho^2}{(\xi^2+\rho^2)^2}\dot{Z} + \frac{\rho^2}{(\xi^2+\rho^2)^2}\left(\frac{\chi^a x^a}{2} + \frac{\dot{\rho}z}{\rho}\right)\right] \tag{4.1.40}$$

The solutoins (4.1.39) and (4.1.40) were ignored when obtaining the mass formula because they are subleading in the $1/\lambda$ expansion but kept here because they are leading contributions to the $U(1)$ current.

So far we have considered solutions valid in the $\xi \ll 1$ region. Now we consider how to find solutions in the region of $1 \ll \xi$. The key observation is that all components of the gauge field are suppressed in the $\rho \ll \xi \ll 1$ region in the large λ limit. This implies that the nonlinear terms in the equations of motion can be neglected. Thus, our strategy is to find a solution to the linearized equation of motion in the $\rho \ll \xi$ region and smoothly connect it to the previous solution in the overlapping region $\rho \ll \xi \ll 1$ (much larger than ρ but much smaller than 1). At $\rho \ll \xi \ll 1$, considering

$$\frac{1}{8\pi^2 a\lambda}\frac{\xi^2+2\rho^2}{(\xi^2+\rho^2)^2} \rightarrow \frac{1}{2a\lambda}\frac{1}{4\pi^2}\frac{1}{\xi^2} \tag{4.1.41}$$

the gauge field can be approximated as follows;

$$\hat{A}_0 \simeq -\frac{1}{2a\lambda}G^{\text{flat}}(\vec{x}, z; \vec{X}, Z) \tag{4.1.42}$$

$$\hat{A}_i \simeq \frac{1}{2a\lambda}\left[\dot{X}^i + \frac{\rho^2}{2}\left\{\frac{\chi^a}{2}\left(\epsilon^{iaj}\frac{\partial}{\partial X^j} - \delta^{ia}\frac{\partial}{\partial Z}\right) + \frac{\dot{\rho}}{\rho}\frac{\partial}{\partial X^i}\right\}\right]G^{\text{flat}}(\vec{x}, z; \vec{X}, Z) \tag{4.1.43}$$

$$\hat{A}_z \simeq \frac{1}{2a\lambda}\left[\dot{Z} + \frac{\rho^2}{2}\left(\frac{\chi^a}{2}\frac{\partial}{\partial X^a} + \frac{\dot{\rho}}{\rho}\frac{\partial}{\partial Z}\right)\right]G^{\text{flat}}(\vec{x}, z; \vec{X}, Z) \tag{4.1.44}$$

$$\begin{aligned} A_0^G \simeq\ & 4\pi^2\rho^2 i W\dot{W}^{-1}G^{\text{flat}}(\vec{x}, z; \vec{X}, Z) \\ & +2\pi^2\rho^2 W\tau^a W^{-1}\left(\dot{X}^i\left(\epsilon_{iaj}\frac{\partial}{\partial X^j} - \delta^{ai}\frac{\partial}{\partial Z}\right) + \dot{Z}\frac{\partial}{\partial X^a}\right)G^{\text{flat}}(\vec{x}, z; \vec{X}, Z) \end{aligned} \tag{4.1.45}$$

$$A_i^G \simeq 2\pi^2\rho^2\left(W\tau^i W^{-1}\frac{\partial}{\partial Z} + \epsilon_{ija}W\tau^a W^{-1}\frac{\partial}{\partial X^j}\right)G^{\text{flat}}(\vec{x}, z; \vec{X}, Z) \tag{4.1.46}$$

$$A_z^G \simeq -2\pi^2\rho^2 W\tau^a W^{-1}\frac{\partial}{\partial X^a}G^{\text{flat}}(\vec{x}, z; \vec{X}, Z), \tag{4.1.47}$$

where the function,

$$G^{\text{flat}}(\vec{x}, z; \vec{X}, Z) = -\frac{1}{4\pi^2\xi^2} \tag{4.1.48}$$

is the Green's function in the flat $\mathbb{R}^4$, which satisfy

$$\partial_M \partial^M G^{\text{flat}}(\vec{x}, z; \vec{X}, Z) = \delta^3(\vec{x} - \vec{X})\delta(z - Z). \tag{4.1.49}$$

We can verify that, as expected, these gauge configurations also satisfy the non-sourced and linearized YM equations and the gauge condition;

$$\partial_\beta \hat{F}^{\alpha\beta} = \partial_\beta F^{\alpha\beta}|_{\text{linear}} = 0 \tag{4.1.50}$$

$$\partial^\alpha \hat{A}_\alpha = 0, \quad \partial^\alpha A^G_\alpha = 0. \tag{4.1.51}$$

In order to connect the solution to the large ξ region, we have to consider the effect of the curved background. In this region, we only need to generalize (4.1.50) and (4.1.51) to the case of nontrivial warp factors as follows,

$$h(z)\partial_\mu^2 \hat{A}_i + \partial_z(k(z)\partial_z \hat{A}_i) = 0, \quad \partial_\mu^2 \hat{A}_z + \partial_z(h(z)^{-1}\partial_z(k(z)\hat{A}_z)) = 0, \tag{4.1.52}$$

$$h(z)\partial_\mu^2 A_i^G + \partial_z(k(z)\partial_z A_i^G) = 0, \quad \partial_\mu^2 A_z^G + \partial_z(h(z)^{-1}\partial_z(k(z)A_z^G)) = 0, \tag{4.1.53}$$

$$h(z)\partial^\mu \hat{A}_\mu + \partial_z(k(z)\hat{A}_z) = 0, \quad h(z)\partial^\mu A_\mu^G + \partial_z(k(z)A_z^G) = 0, \tag{4.1.54}$$

where, in this case, all gauge field components are suppressed again, so the nonlinear terms can be neglected. To solve this, we need to define two Green's functuons in the curved space,

$$G(\vec{x}, z; \vec{X}, Z) = \kappa \sum_{n=1}^{\infty} \psi_n(z)\psi_n(Z)Y_n(|\vec{x} - \vec{X}|) \tag{4.1.55}$$

$$H(\vec{x}, z; \vec{X}, Z) = \kappa \sum_{n=0}^{\infty} \phi_n(z)\phi_n(Z)Y_n(|\vec{x} - \vec{X}|), \tag{4.1.56}$$

where the $\psi_n(z)$ and $\phi_n(z)$ are the complete sets defined in the mode expansion, and $Y_n(r)$ is the Yukawa potential with meson mass $m_n = \sqrt{\lambda_n}$,

$$Y_n(r) = -\frac{1}{4\pi}\frac{e^{-\sqrt{\lambda_n}r}}{r}, \tag{4.1.57}$$

which satisfy

$$(\partial_i^2 - \lambda_n)Y_n(|\vec{x} - \vec{X}|) = \delta^3(\vec{x} - \vec{X}). \tag{4.1.58}$$

Here, it is no coincidence that the complete sets of mode expansions were used here. Recalling that the discussion of mode expansion also dropped the nonlinear terms (Sect. 2.3.2), it is easy to see that (4.1.52) and (4.1.53) is, for instance,

$$h(z)\partial_\mu^2(\psi_n Y_n) + \partial_z(k(z)\partial_z(\psi_n Y_n)) = 0 \tag{4.1.59}$$

$$\to \psi_n \partial_\mu^2 Y_n + h(z)^{-1}\partial_z(k(z)\partial_z\psi_n)Y_n = 0 \tag{4.1.60}$$

$$\to \psi_n(\partial_\mu^2 - \lambda_n)Y_n = 0. \tag{4.1.61}$$

Using (2.3.17) and (4.1.58), one can easily verify

$$h(z)\partial_i^2 G + \partial_z(k(z)\partial_z G) = \delta^3(\vec{x} - \vec{X})\delta(z - Z), \tag{4.1.62}$$

$$\partial_i^2 H + \partial_z(h(z)^{-1}\partial_z(k(z)H)) = k(z)^{-1}\delta^3(\vec{x} - \vec{X})\delta(z - Z), \tag{4.1.63}$$

$$\partial_z(k(z)H) + h(z)\partial_Z G = 0, \tag{4.1.64}$$

where we used the condition

$$\kappa h(z)\sum_{n=1}^{\infty}\psi_n(z)\psi_n(Z) = \delta(z - Z), \quad \kappa k(z)\sum_{n=1}^{\infty}\phi_n(z)\phi_n(Z) = \delta(z - Z) \tag{4.1.65}$$

for the complete system. This condition ensure normalized orthogonality of the eigenfunctions for ψ_m and ϕ_m, as follow,

$$\begin{aligned}\int dz\psi_m(z)\delta(z - Z) = \psi_m(Z) &= \sum_{n=1}^{\infty}\int dz\kappa h(z)\psi_m(z)\psi_n(z)\psi_n(Z)\\ &= \sum_{n=1}^{\infty}\delta_{nm}\psi_n(Z)\end{aligned} \tag{4.1.66}$$

$$\begin{aligned}\int dz\phi_m(z)\delta(z - Z) = \phi_m(Z) &= \sum_{n=1}^{\infty}\kappa\int dzk(z)\phi_m(z)\phi_n(z)\phi_n(Z)\\ &= \sum_{n=1}^{\infty}\delta_{nm}\phi_n(Z).\end{aligned} \tag{4.1.67}$$

Thus, the completeness condition (4.1.65) can also be used in the case where n is a sum of even numbers or odd numbers only, where we will use these conditions later. From the above, the solution is given as

$$\hat{A}_0 = -\frac{1}{2a\lambda} G\left(\vec{x}, z; \vec{X}, Z\right),$$
$$\hat{A}_i = \frac{1}{2a\lambda}\left[\dot{X}^i + \frac{\rho^2}{2}\left(\frac{\chi^a}{2}\left(\epsilon^{iaj}\frac{\partial}{\partial X^j} - \delta^{ia}\frac{\partial}{\partial Z}\right) + \frac{\dot{\rho}}{\rho}\frac{\partial}{\partial X^i}\right)\right] G\left(\vec{x}, z; \vec{X}, Z\right),$$
$$\hat{A}_z = \frac{1}{2a\lambda}\left[\dot{Z} + \frac{\rho^2}{2}\left(\frac{\chi^a}{2}\frac{\partial}{\partial X^a} + \frac{\dot{\rho}}{\rho}\frac{\partial}{\partial Z}\right)\right] H\left(\vec{x}, z; \vec{X}, Z\right), \tag{4.1.68}$$
$$A_0 = 4\pi^2\rho^2 i W \dot{W}^{-1} G\left(\vec{x}, z; \vec{X}, Z\right)$$
$$+2\pi^2\rho^2 W\tau^a W^{-1}\left(\dot{X}^i\left(\epsilon^{iaj}\frac{\partial}{\partial X^j} - \delta^{ia}\frac{\partial}{\partial Z}\right) + \dot{Z}\frac{\partial}{\partial X^a}\right) G\left(\vec{x}, z; \vec{X}, Z\right),$$
$$A_i = -2\pi^2\rho^2 W\tau^a W^{-1}\left(\epsilon^{iaj}\frac{\partial}{\partial X^j} - \delta^{ia}\frac{\partial}{\partial Z}\right) G\left(\vec{x}, z; \vec{X}, Z\right),$$
$$A_z = -2\pi^2\rho^2 W\tau^a W^{-1}\frac{\partial}{\partial X^a} H\left(\vec{x}, z; \vec{X}, Z\right), \tag{4.1.69}$$

where the index i runs $1-3$.

4.1.2.3 The Chiral Current

We substitute the classical solution into the definition of above chiral current. In the following calculations, since we consider wave functions in low-lying states, Z can be considered as $Z \sim \mathcal{O}(\lambda^{-1/2}N_c^{-1/2}) \ll 1$, from which we can approximate that $h(Z) = k(Z) = 1$. The eigenvalue equation satisfied by the complete sets $\psi_n(z)$ of mode expansions is then approximated by

$$-\partial_Z^2\psi_n(Z) \simeq \lambda_n\psi_n(Z). \tag{4.1.70}$$

Using this and (4.1.64), the following approximation holds;

$$\partial_Z H + \partial_z G \simeq 0,\ (\partial_i^2 + \partial_Z^2)G \simeq 0,\ (\partial_i^2 + \partial_Z^2)H \simeq 0 \tag{4.1.71}$$

By taking this approximation into account, the field strength is given as

$$\hat{F}_{iz} \simeq \frac{1}{2a\lambda}\left[\dot{Z}\partial_i H - \dot{X}^i\partial_z G - \frac{\rho^2\chi^a}{4}((\partial_i\partial_a - \delta^{ia}\partial_j^2)H - \epsilon^{iaj}\partial_j\partial_z G)\right] \tag{4.1.72}$$
$$F_{0z} \simeq 2\pi^2\partial_0(\rho^2 W\tau^a W^{-1})\partial_a H - 4\pi^2\rho^2 i W\dot{W}^{-1}\partial_z G$$
$$-2\pi^2\rho^2 W\tau^a W^{-1}\dot{X}^i((\partial_i\partial_a - \delta^{ia}\partial_j^2)H - \epsilon^{iaj}\partial_j\partial_z G) \tag{4.1.73}$$
$$F_{iz} \simeq 2\pi^2\rho^2 W\tau^a W^{-1}((\partial_i\partial_a - \delta^{ia}\partial_j^2)H - \epsilon^{iaj}\partial_j\partial_z G). \tag{4.1.74}$$

With (4.1.26), $G^{V/A}(Z, r)$, $H^{V/A}(Z, r)$ are defined by

$$G^V(Z,r) = \big[k(z)\partial_z G\big]_{z=-\infty}^{z=+\infty} = -\sum_{n=1}^{\infty} g_{v^n}\psi_{2n-1}(Z)Y_{2n-1}(r) \tag{4.1.75}$$

$$G^A(Z,r) = \big[\psi_0(z)k(z)\partial_z G\big]_{z=-\infty}^{z=+\infty} = -\sum_{n=1}^{\infty} g_{a^n}\psi_{2n}(Z)Y_{2n}(r) \tag{4.1.76}$$

$$H^V(Z,r) = \big[k(z)H\big]_{z=-\infty}^{z=+\infty} = -\sum_{n=1}^{\infty} \frac{g_{v^n}}{\lambda_{2n-1}}\partial_Z\psi_{2n-1}(Z)Y_{2n}(r) \tag{4.1.77}$$

$$H^A(Z,r) = \big[\psi_0(z)k(z)H\big]_{z=-\infty}^{z=+\infty} = -\frac{1}{2\pi^2}\frac{1}{k(Z)}\frac{1}{r} - \sum_{n=1}^{\infty} \frac{g_{a^n}}{\lambda_{2n}}\partial_Z\psi_{2n}(Z)Y_{2n}(r). \tag{4.1.78}$$

From these, The vector and the axial vector current are obtained as follow;

$$\hat{J}^0_{V,A} = \frac{N_c}{2}G^{V,A} \tag{4.1.79}$$

$$\hat{J}^i_{V,A} = -\frac{N_c}{2}\Big[\dot{Z}\partial_i H^{V,A} - \dot{X}^i G^{V,A} - \frac{\rho^2\chi^a}{4}\big((\partial_a\partial_i - \delta^{ia}\partial_j^2)H^{V,A} - \epsilon^{iaj}\partial_j G^{V,A}\big)\Big] \tag{4.1.80}$$

$$\begin{aligned} J^0_{V,A} = {} & 2\pi^2\kappa\Big[\partial_0(\rho^2 W\tau^a W^{-1})\partial_a H^{V,A} - 2\rho^2 i W\dot{W}^{-1}G^{V,A} \\ & -\rho^2 W\tau^a W^{-1}\dot{X}^i\big((\partial_a\partial_i - \delta^{ia}\partial_j^2)H^{V,A} - \epsilon^{iaj}\partial_j G^{V,A}\big)\Big] \end{aligned} \tag{4.1.81}$$

$$J^i_{V,A} = -2\pi^2\kappa\rho^2 W\tau^a W^{-1}\big((\partial_a\partial_i - \delta^{ia}\partial_j^2)H^{V,A} - \epsilon^{iaj}\partial_j G^{V,A}\big), \tag{4.1.82}$$

where, since $\psi_{2n-1}(Z)$ are an even function of Z and $\psi_{2n}(Z)$ an odd function of Z, G^V, H^A are an even function and G^A, H^V an odd function.

4.1.3 The Chiral Current in the 5-Dimensional Effective Theory

4.1.3.1 Chiral Symmetry in the Sakai-Sugimoto Model

Now, we define the chiral current as of the Noether current of chiral symmetry from the point of view of the effective theory of hadrons [2]. Therefore, we first discuss the chiral symmetry of this model.

The pion field of this model is defiened by the Atiyah-Manton construction [7] in gauge condition $\mathcal{A}_M(x^\mu, z) \to 0$ $(z \to \pm\infty)$ as follows;

$$U(x^\mu) = \mathrm{P}\exp\Big(-i\int_{-\infty}^{+\infty} dz\mathcal{A}_z(x^\mu, z)\Big), \tag{4.1.83}$$

where the path ordering P is product of left at $z \to +\infty$ to right at $z \to -\infty$. The chiral transformation of the pion field,

$$U(x^\mu) = g_L U(x^\mu) g_R, \qquad (4.1.84)$$
$$(g_L, g_R) \in U(N_f)_L \times U(N_f)_R,$$

is realized by a gauge transformation of the flavor $SU(N_f)$ gauge field as follows;

$$\mathcal{A}_M \to g\mathcal{A}_M g^{-1} - ig\partial_M g^{-1}, \qquad (4.1.85)$$

$$g(x^\mu, z) \to \begin{cases} g_L \; (z \to +\infty) \\ g_R \; (z \to -\infty) \end{cases} \qquad (4.1.86)$$

with $g(x^\mu, z) \in SU(N_f)$ and constants (g_L, g_R).

4.1.3.2 Noether Currents

The infinitesimal local gauge transformation,

$$\delta_\xi \mathcal{A}_M(x^\mu, z) = \epsilon(x^\mu, z)\mathcal{D}_M\zeta(x^\mu, z), \qquad (4.1.87)$$

leads to the following Noether currents;

$$J_\zeta^M = J_{YM\zeta}^M + J_{CS\zeta}^M, \qquad (4.1.88)$$
$$J_{YM\zeta}^\mu(x, z) = -2\kappa \mathrm{tr}\big(h(z)\mathcal{F}^{\mu\nu}\mathcal{D}_\nu\zeta + k(z)\mathcal{F}^{\mu z}\mathcal{D}_z\zeta\big),$$
$$J_{YM\zeta}^z(x, z) = -2\kappa k(z)\mathrm{tr}\big(\mathcal{F}^{z\nu}\mathcal{D}_z\zeta\big),$$
$$J_{CS\zeta}^M(x.z) = -\frac{N_c}{64\pi^2}\epsilon^{MNPQR}\mathrm{tr}\big(\{\mathcal{F}_{NP}, \mathcal{F}_{QR}\}\zeta\big).$$

with $u(N_f)$ Lie algebra $\zeta(x^\mu, z)$, with a function $\epsilon(x^\mu, z)$ vanishing at infinity, and the covariant derivative $\mathcal{D}_M\zeta(x^\mu, z) = \partial_M\zeta + i\mathcal{A}_M\zeta$. Because of the chiral symmetry of the Sakai-Sugimoto model is related to the $SU(N_f)$ gauge transformation, The chiral current in the 4-dimensional space-time is defined as

$$j_\zeta^\mu(x) = \int_{-\infty}^{+\infty} dz J_\zeta^\mu(x, z). \qquad (4.1.89)$$

Here, to satisfy the 4-dimensional current conservation law, we impose the following boundary condition,

$$J_\zeta^z(x, z \to \pm\infty) = 0. \qquad (4.1.90)$$

With unit and Pauli matrices $t_C = 1/2(I, \tau^a)$ $(C = 0, 1, 2, 3)$ and

$$\psi_{\pm}(z) = \frac{1}{2} \pm \frac{1}{\pi}\mathrm{arctan}z \rightarrow \begin{cases} 1 \ (z \rightarrow \pm\infty) \\ 0 \ (z \rightarrow \mp\infty) \end{cases}, \tag{4.1.91}$$

we adopt as ζ the following one:

$$\zeta(x, z) = \psi_{\pm}(z) t_C, \tag{4.1.92}$$

then the expression of our current (4.1.89) leads to left/right current $j^{\mu}_{L/R}(x)$. Therefore, the vector/axial-vector current is then defined as follows;

$$j_V^{\mu,C} = j_L^{\mu,C} + j_R^{\mu,C}, \tag{4.1.93}$$

$$j_A^{\mu,C} = j_L^{\mu,C} - j_R^{\mu,C}. \tag{4.1.94}$$

$$\left(j^{\mu}_{L/R}(x) = j^{\mu,C}_{L/R} t_C = j^{\mu,a}_{L/R}\frac{\tau^a}{2} + \hat{j}^{\mu}_{L/R}\frac{I}{2} \right) \tag{4.1.95}$$

Here, we define

$$\psi_V(z) = \psi_+(z) + \psi_-(z) = 1 \tag{4.1.96}$$

$$\psi_A(z) = \psi_+(z) - \psi_-(z) = \frac{2}{\pi}\mathrm{arctan}z, \tag{4.1.97}$$

then the vector/axial vector current are written by

$$\begin{aligned} J^{\mu}_{V/A,a} = & -2i\kappa\mathrm{tr}\left\{(h(z)[\mathcal{F}^{\mu z}, \mathcal{A}_\nu] + k(z)[\mathcal{F}^{\mu z}, \mathcal{A}_z])t_a\right\}\psi_{V/A}(z) \\ & -2\kappa k(z)\mathrm{tr}(\mathcal{F}^{\mu z}t_a)\frac{d\psi_{V/A}(z)}{dz} \\ & -\frac{N_c}{64\pi^2}\epsilon^{\mu NPQR}\mathrm{tr}\left(\{\mathcal{F}_{NP}, \mathcal{F}_{QR}\}t_a\right)\psi_{V/A}(z). \end{aligned} \tag{4.1.98}$$

Because of the $A^{\mathrm{cl}}_{i,z} = 0$ in the static case, the baryon number current become

$$\begin{aligned} j^0_B(x) = & \frac{2}{N_c}\hat{j}^0_V(x) = \int_{-\infty}^{\infty} dz \hat{J}^0_V(x) \\ = & -\frac{N_c}{64\pi^2}\epsilon^{\mu NPQR}\int_{-\infty}^{\infty} dz \mathrm{tr}(F_{NP}F_{QR}), \end{aligned} \tag{4.1.99}$$

which identifies with the topological number current of instanton solutions.

4.1.3.3 The $1/\lambda$ Expansion of the Current

In the following, we will carry on the discussion by using the $1/\lambda$ expansion order by order. Therefore, we will consider rescaling the current by λ. With $\tilde{\psi}_{V,A}(z)$ and $\kappa = \mathcal{O}(\lambda)$, by using the following rescaling,

$$x^M \to \lambda^{-1/2}x^M, \quad \mathcal{A}_M \to \lambda^{1/2}\mathcal{A}_M$$
$$\mathcal{F}_{MN} \to \lambda\mathcal{F}_{MN}, \quad \mathcal{F}_{0M} \to \lambda^{1/2}\mathcal{F}_{0M}, \tag{4.1.100}$$

the chiral current are written as

$$\begin{aligned} J^0_{V/A,a}(x,z) = & -2i\kappa\lambda\Big\{\mathrm{tr}\Big(\sum_{M=i,z}[\mathcal{F}_{0M},\mathcal{A}_M]t_a\Big)\tilde{\psi}_{V,A}(z) + \mathrm{tr}(\mathcal{F}_{0z}t_a)\frac{d\tilde{\psi}_{V,A}(z)}{dz}\Big\} \\ & -\frac{N_c}{64\pi^2}\lambda^2\epsilon_{MNPQ}\mathrm{tr}(\{\mathcal{F}_{MN},\mathcal{F}_{PQ}\}t_a)\tilde{\psi}_{V,A}(z) \\ & -2\kappa\Big\{\mathrm{tr}\Big(\Big(-\frac{z^2}{3}\sum_j[\mathcal{F}_{0j},\mathcal{A}_j] + z^2[\mathcal{F}_{0z},\mathcal{A}_z]\Big)t_a\Big)\tilde{\psi}_{V,A}(z) + z^2\mathrm{tr}(\mathcal{F}_{iz}t_a)\frac{d\tilde{\psi}_{V,A}(z)}{dz}\Big\} \end{aligned} \tag{4.1.101}$$

$$\begin{aligned} J^i_{V/A,a}(x,z) = & -2\kappa\lambda^{3/2}\Big\{\mathrm{tr}\Big(\sum_{M=i,z}[\mathcal{F}_{iM},\mathcal{A}_M]t_a\Big)\tilde{\psi}_{V,A}(z) + \mathrm{tr}(\mathcal{F}_{iz}t_a)\frac{d\tilde{\psi}_{V,A}(z)}{dz}\Big\} \\ & -2\kappa\lambda^{1/2}\Big\{\mathrm{tr}\Big(\Big(-\frac{z^2}{3}\sum_j[\mathcal{F}_{ij},\mathcal{A}_j] + z^2[\mathcal{F}_{iz},\mathcal{A}_z]\Big)t_a\Big)\tilde{\psi}_{V,A}(z) + z^2\mathrm{tr}(\mathcal{F}_{iz}t_a)\frac{d\tilde{\psi}_{V,A}(z)}{dz}\Big\} \\ & +\frac{N_c\lambda^{3/2}}{8\pi^2}\epsilon_{ijk}\mathrm{tr}\big[(\mathcal{F}_{jk}\mathcal{F}_{0z} + 2\mathcal{F}_{0j}\mathcal{F}_{kz})t_a\big]\tilde{\psi}_{V,A}(z), \end{aligned} \tag{4.1.102}$$

where the gauge fields (4.1.33), (4.1.34), (4.1.38), (4.1.38) and (4.1.38) rescaled as follows;

$$A^G_0 = -i(1-f(\xi))W\dot{W}^{-1} + i(1-f(\xi))\dot{X}^M W(g^{-1}\partial_M g)W^{-1} \tag{4.1.103}$$
$$A^G_M = -i(1-f(\xi))W(g^{-1}\partial_M g)W^{-1} \tag{4.1.104}$$
$$\hat{A}_0 = \frac{1}{8\pi^2 a}\frac{\xi^2+2\rho^2}{(\xi^2+\rho^2)^2} \tag{4.1.105}$$
$$\hat{A}_i = -\frac{1}{8\pi^2 a}\Big[\frac{\xi^2+2\rho^2}{(\xi^2+\rho^2)^2}\dot{X}^i + \frac{\rho^2}{(\xi^2+\rho^2)^2}\Big(\frac{\chi^a}{2}(\epsilon^{iaj}x^j - \delta^{ia}z) + \frac{\dot{\rho}x^i}{\rho}\Big)\Big] \tag{4.1.106}$$
$$\hat{A}_z = -\frac{1}{8\pi^2 a}\Big[\frac{\xi^2+2\rho^2}{(\xi^2+\rho^2)^2}\dot{Z} + \frac{\rho^2}{(\xi^2+\rho^2)^2}\Big(\frac{\chi^a x^a}{2} + \frac{\dot{\rho}z}{\rho}\Big)\Big]. \tag{4.1.107}$$

In the following calculations, we will proceed with the analysis by considering only terms that give the main contribution to the $1/\lambda$ expansion for a certain physical quantity. For some physical quantities, the contribution of the first term as leading vanishes, hence the second and third terms of $J^M_{V/A,a}$, which are subleading of the $1/\lambda$ expansion, provide the main contribution.

We give some comments on what we learn from the above rescaling for later discussion. First, we find that the effect of the warp factor is a contribution from subleading in both the time and spatial components of the current. Therefore, unless the contribution from the leading term is zero, we can treat it as $k(z) = h(z) \sim 1$. Next, regarding the contribution from the CS term, on the one hand, it contributes to the leading term of the time component of the current. On the other hand, it gives a subleading contribution to the spatial component. The spatial component of the U(1) gauge field can be set zero in the calculation of leading order, since it only contributes to the CS part of the current. These facts are employed in the following calculations.

4.1.4 Some Comments for Two Definitions of Current

4.1.4.1 The Relation Between the Two Current Definitions

First, we discuss the connection between the definitions of the two currents. The 5d Noether current [2] is transformed as follows,

$$\begin{aligned} J^{\mu}_{\zeta}(x,z) = & -2\kappa\partial_{\nu}\mathrm{tr}\big[h(z)\mathcal{F}^{\mu\nu}(x,z)\zeta(x,z)\big] - 2\kappa\partial_{z}\mathrm{tr}\big[h(z)\mathcal{F}^{\mu z}(x,z)\zeta(x,z)\big] \\ & -\mathrm{tr}\big[(\text{EOM term}) \times \zeta(x,z)\big]. \end{aligned} \tag{4.1.108}$$

With the current $\tilde{j}^{\mu}_{L/R}(x)$ based on the bulk/boundary correspondenc, $j^{\mu}_{\zeta}(x) = \int dz J^{\mu}_{\zeta}(x,z)$ is written by

$$j^{\mu}_{\zeta}(x) = \mathrm{tr}(\zeta_L \tilde{j}^{\mu}_L(x) + \zeta_R \tilde{j}^{\mu}_R(x)) + \partial_{\nu}\chi^{\mu\nu}(x) + (\text{EOM term}), \tag{4.1.109}$$

where $\chi^{\mu\nu}(x)$ is the following antisymmetric tensor,

$$\chi^{\mu\nu} = -\chi^{\nu\mu} = -2\kappa \int_{-\infty}^{\infty} dz \mathrm{tr}\big[h(z)\mathcal{F}^{\mu\nu}(x,z)\zeta(x,z)\big]. \tag{4.1.110}$$

We will remark on what can be learned from the above relation. First, the identity $\partial_{\mu}\partial_{\nu}\chi^{\mu\nu} = -\partial_{\mu}\partial_{\nu}\chi^{\mu\nu} = 0$ indicates that the difference between the two currents does not contradict the conservation law of currents. Also, if the surface term coming from $\partial_i\chi^{0i}$ can be ignored, then we can show that the conserved charge, such as the baryon number defined by $Q = \int d^3x j^0(x)$, leads to the same result with the two different definitions. Similarly, global quantities such as axial vector coupling, which involve spatial integration of currents, do not depend on the definition of the two currents. However, since local currents take different forms from each other, the two currents lead to different results for physical quantities, for example, the isovector magnetic moment $\boldsymbol{\mu}_{I=1}$ (4.2.27) and isospin charge density, to which the local currents contribute explicitly.

4.1.4.2 Gauge Non-invariance of the Current

Next, while the current $\tilde{j}^{\mu}_{L/R}(x)$ based on the bulk/boundary correspondence is gauge-invariant from the above expressions, the 5d Noether chiral current is generally gauge-dependent. As a result, it is gauge-dependent concerning physical quantities that depend on the local current form. It is still unknown how this problem should be solved. Here, conserved charges, such as baryon number, take the same form in both definitions, as described above. Therefore, these quantities can still be defined as gauge invariant. Related to this, there is another problem that $\zeta(x, z)$ cannot be uniquely determined. The quantity $\zeta(x, z)$ only needs to be any function that satisfies the boundary condition (4.1.86) and takes values on $U(N_f)$. Thus, using this non-uniqueness, we see that the current is gauge-invariant under a simultaneous gauge transformation

$$\begin{aligned} \mathcal{A}_M &\to g\mathcal{A}_M g^{-1} - ig\partial_M g^{-1} \\ \zeta &\to g\zeta g^{-1} \end{aligned} \tag{4.1.111}$$

In this thesis, we have already mentioned that we determine this $\zeta(x, z)$ to reduce to the chiral current of the Skyrme model by dimensionality reduction. Considering that the hadron effective action of the Sakai-Sugimoto model has the same structure as that of the phenomenological model using hidden local symmetry proposed by Son-Stephanov [3], the above treatment seems reasonable. However, it cannot be claimed that this method follows the dictionary of gauge/gravity (string) correspondence.

4.2 The Static Properties of the Nucleon

Using currents and classical solutions obtained above, in this section, we review the analysis of static properties of nucleons. We have introduced two definitions of currents and summarized the results obtained from each current. Note that the choice of parameters is different for each analysis of the two currents. When using the chiral current obtained from the GKP-Witten method (hereafter named GKP-Witten current), we have chosen $(\kappa, M_{KK}) = (0.00745, 940\ \text{MeV})$ to reproduce the ρ meson mass (776 MeV) and the pion decay constant (92.4 MeV) following Ref. [1]. On the other hand, when using the chiral current as a 5d Noether current (called 5d Noether current), we consider the nucleon mass as the classical mass M_0 of the soliton according to Ref. [2], and choose a parameter set $(\kappa, M_{KK}) = (0.0243, 488\ \text{MeV})$ that reproduces the mass of nucleon (939 MeV) and the mass splitting of $N(940)$ and $\Delta(1232)$ (293 MeV).

4.2.1 *Baryon Number Current and Isoscalar Mean Square Radius*

We have considered baryons as instantons in this doctoral thesis. Therefore, the baryon number corresponds to the topological number of instantons. Since the topological number current is defined as

$$J_B^0(x) = \frac{1}{32\pi^2}\epsilon_{ONPQR}\int_{-\infty}^{\infty} dz \mathrm{tr}(\mathcal{F}_{NP}\mathcal{F}_{QR}), \tag{4.2.1}$$

we therefore expect the baryon number current to be defined in the same form. The baryon number current obtained from the GKP-Wittten current coincides with the instanton topological number current, as shown in (4.1.18). A similar fact can be derived from the baryon number currents obtained from the Neother currents, as shown in (4.1.99).

The baryon number density is defined by

$$\rho_B(r) = 4\pi r^2 \left\langle J_B^0(r)\right\rangle. \tag{4.2.2}$$

Using this, the isoscalar mean square radius is written by

$$\left\langle r^2\right\rangle_{I=0} = \int_0^{\infty} dr r^2 \rho_B(r). \tag{4.2.3}$$

In the following, we will calculate the baryon number, baryon number density, and isoscalar mean square radius for each current.

■ GKP-Witten current

We confirm that the baryon number $B = 1$ is actually obtained by substituting the classical solution into the baryon number currents. First, by substituting the asymptotic solution (4.1.69) and (4.1.68) into the GKP-Witten current, we obtain

$$J_B^0 = -\sum_{n=1}^{\infty} g_{v^n}\psi_{2n-1}(Z)Y_{2n-1}(r). \tag{4.2.4}$$

The baryon number is computed as follows,

$$\begin{aligned} N_B &= \sum_{n=1}^{\infty} g_{v^n}\left\langle\psi_{2n-1}(Z)\right\rangle\left(-\int_0^{\infty} dr 4\pi r^2 Y_{2n-1}(r)\right) = \sum_{n=1}^{\infty}\frac{g_{v^n}}{\lambda_{2n-1}}\left\langle\psi_{2n-1}(Z)\right\rangle \\ &= \sum_{n=1}^{\infty}\kappa\int_{-\infty}^{\infty} dz h(z)\psi_{2n-1}(z)\left\langle\psi_{2n-1}(Z)\right\rangle = \left\langle\int_{-\infty}^{\infty} dz\delta(z-Z)\right\rangle = 1, \end{aligned} \tag{4.2.5}$$

where the shorthand notation $\langle\mathcal{O}\rangle = \langle B|\mathcal{O}|B\rangle$ is used hereafter, with attention to normalization, which is defined as follows;

$$\langle\psi_{2n-1}(Z)\rangle = \langle n_Z|\psi_{2n-1}(Z)|n_Z\rangle \tag{4.2.6}$$

by using (2.4.123).

The baryon number density is

$$\rho_B(r) = r\sum_{n=1}^{\infty} g_{v^n}\,\langle\psi_{2n-1}(Z)\rangle\, e^{-\sqrt{\lambda_{2n-1}}}r, \tag{4.2.7}$$

therefore, the isoscalar mean square radius is defined by

$$\langle r^2\rangle_{I=0} = \int_0^{\infty} drr^2\rho_B(r) = 6\sum_{n=1}^{\infty}\frac{g_{v^n}}{\lambda_{2n-1}^2}\,\langle\psi_{2n-1}(Z)\rangle \tag{4.2.8}$$

Numerical calculations show that this value is estimated to be

$$\sqrt{\langle r^2\rangle_{I=0}} \simeq 0.785\text{fm}, \tag{4.2.9}$$

which roughly reproduces the experimental value of 0.806 fm.

■ The 5d Noether current

Substituting the classical BPST instanton solution for the 5d Noether current, we obtain

$$\begin{aligned} j_B^0(x,z) &= \int_{-\infty}^{\infty}\frac{2}{N_c}J_{V,0}^0(x,z) = -\frac{1}{32\pi^2}\epsilon_{MNPQ}\int_{-\infty}^{\infty}\text{tr}(\{F_{MN},F_{PQ}\}\frac{I}{2}) \\ &= -\frac{4}{32\pi^2}\epsilon_{ijk}\int_{-\infty}^{\infty}\text{tr}(F_{ij}F_{kz}) = \frac{15}{8\pi}\frac{\rho^4}{(r^2+\rho^2)^{7/2}}, \end{aligned} \tag{4.2.10}$$

where the contribution from $\hat{F}_{ij}$, $\hat{F}_{iz}$ is negligible here, because the spatial integration leads to zero. We also dropped the terms involving $\dot{Z}$ in order to take the expectation value with the nucleon state. Furthermore, the effect of the warp factor is treated as $h(z) \sim k(z) \sim 1$ because it is a contribution of sublearding in the $1/\lambda$ expansion. If we spatial integrate this, we get

$$N_B = \int_0^{\infty} 4\pi r^2 dr\frac{15}{8\pi}\frac{\rho^4}{(r^2+\rho^2)^{7/2}} = 1. \tag{4.2.11}$$

By the same calculation as (4.2.8), the isoscalar mean square radius is evaluated to be

$$\sqrt{\langle r^2\rangle_{I=0}} = \sqrt{\frac{3\langle\rho^2\rangle}{2}} = 0.82\text{fm}, \tag{4.2.12}$$

which is larger than the result by GKP-Witten current. Here, the expected value of ρ^2 can be obtained analytically as follows;

$$\langle\rho\rangle = \langle n_\rho|\rho^2|n_\rho\rangle = \frac{\int_0^\infty \rho^3 d\rho\rho^2\rho^{-2+4\sqrt{1+N_c^2/5}}e^{-\frac{2M_0}{\sqrt{6}}\rho^2}}{\int_0^\infty \rho^3 d\rho\rho^{-2+4\sqrt{1+N_c^2/5}}e^{-\frac{2M_0}{\sqrt{6}}\rho^2}} = \frac{\sqrt{6}}{16\pi^2\kappa}\left(1+2\sqrt{1+\frac{N_c^2}{5}}\right). \tag{4.2.13}$$

by using (2.4.121).

4.2.2 *Charge Density*

In this subsection, we discuss the nucleon charge density, which is written as

$$J^0_{\text{em}}(t,x) = J^0_{V,a=3}(t,x) + \frac{1}{2}J^0_B(t,x), \tag{4.2.14}$$

with the isospin density $J^0_{V.a=3}(t,x)$ and the baryon number current. Therefore, the electric charge is calculated as

$$Q_{\text{em}} = \int d^3x J^0_{\text{em}}(t,x). \tag{4.2.15}$$

As a result, $Q_{\text{em}} = I_3 + N_B/2$ should be obtained. Also, the electric charge radius is given by

$$\langle r^2\rangle_E = \int d^3x r^2 J^0_{\text{em}}(t,x). \tag{4.2.16}$$

■ GKP-Witten current

Since we take the expectation value with the nucleon state, we only need to consider the even function terms of Z in the current $J^0_{\text{em}}(t,x)$. Therefore, the current matrix element is

$$\langle J^0_{\text{em}}\rangle = \left(i4\pi^2\kappa\rho^2\text{tr}(\tau^3 W\dot{W}^{-1}) - \frac{1}{2}\right)\sum_{n=1}^{\infty} g_{v^n}\langle\psi_{2n-1}(Z)\rangle\int_0^\infty 4\pi r^2 Y_{2n-1}(r) = I_3 + \frac{N_B}{2} = I_3 + \frac{1}{2}, \tag{4.2.17}$$

where $I_a = -i4\pi^2\kappa\rho^2\text{tr}(\tau^a W\dot{W}^{-1})$ and the same calculation as (4.2.5) was performed. Thus, as expected, $Q_{\text{em}} = 1$ for the proton case and $Q_{\text{em}} = 0$ for the neutron case. You can see from the calculation of (4.2.17) that we can check $\langle r^2\rangle_{I=0} = \langle r^2\rangle_{I=1}$.

Therefore, the electric charge radius is obtained as

$$\langle r^2\rangle_{E,p} = \langle r^2\rangle_{I=0} = (0.785\ \text{fm})^2\ \text{(proton)} \tag{4.2.18}$$

$$\langle r^2\rangle_{E,n} = 0 \qquad \text{(neutron)}, \tag{4.2.19}$$

where the experimental values are $\langle r^2\rangle^{\text{exp}}_{E,p} \simeq (0.875\ \text{fm})^2$, $\langle r^2\rangle^{\text{exp}}_{E,n} \simeq -0.116\ \text{fm}^2$, which fails to reproduce neutron data.

■ The 5d Noether current

The electric charge radius can be written as the sum of the isoscalar mean square radius and the isovector mean square radius discussed in the previous subsection. Therefore, here we discuss the isovector mean square radius before considering the electric charge radius. The five-dimensional isovector current is written by

$$J^0_{V,a}(x,z) = 2\kappa \text{tr}\Big[\Big(\sum_{M=i,z} i\big[\frac{\partial}{\partial\rho}A^{\text{cl}}_M, A^{\text{cl}}_M\big]\dot{\rho}(t) - \sum_{M=i,z} i\big[D^{\text{cl}}_M\overline{\Phi}_b, A^{\text{cl}}_M\big]\chi^b(t)\Big)W(t)^{-1}t_a W(t)\Big]. \tag{4.2.20}$$

By using

$$\sum_{M=i,z} i\big[D^{\text{cl}}_M\overline{\Phi}_b, A^{\text{cl}}_M\big] = \frac{8}{\rho^2}\tilde{f}(\xi)^3 t_a, \tag{4.2.21}$$

the isovector current is calculated

$$j^0_{V,a}(t,x) = \int_{-\infty}^{\infty} dz J^0_{V,a}(x,z) = \frac{3}{4\pi}\frac{\rho^2}{(r^2+\rho^2)^{5/2}} I_a. \tag{4.2.22}$$

Therefore, the isovector mean square radius is defined by

$$\langle r^2\rangle_{I=1} = \frac{\langle\int d^3x r^2 j^0_{V,a=3}(t,x)\rangle_\rho}{\langle\int d^3x j^0_{V,a=3}(t,x)\rangle_\rho}, \tag{4.2.23}$$

which is logarithmically divergent at $r \to \infty$. This occurs in the Skyrme model as well, and it is quite natural to obtain similar results from this current, which is determined to realize the chiral current of the Skyrme model with a low energy limit.

4.2.3 Magnetic Moment

The magnetic moment $\boldsymbol{\mu}$ is defined as

$$\boldsymbol{\mu} = \frac{1}{2}\int d^3x \boldsymbol{x} \times \boldsymbol{J}_{\text{em}}, \tag{4.2.24}$$

where $\boldsymbol{J}_{\text{em}} = (J^1_{\text{em}}, J^2_{\text{em}}, J^3_{\text{em}})$ is the electromagnetic current

$$\boldsymbol{J}_{\text{em}} = \boldsymbol{J}_{V,a=3} + \frac{1}{2}\boldsymbol{J}_B. \tag{4.2.25}$$

The magnetic moments are decomposed into isovector and isoscalar magnetic moments, respectively, as follows,

$$\boldsymbol{\mu} = \frac{1}{2}\boldsymbol{\mu}_{I=1} + \frac{1}{2}\boldsymbol{\mu}_{I=0} \tag{4.2.26}$$

$$\boldsymbol{\mu}_{I=1} = \int d^3x \boldsymbol{x} \times \boldsymbol{J}_{V,a} \tag{4.2.27}$$

$$\boldsymbol{\mu}_{I=0} = \frac{1}{2}\int d^3x \boldsymbol{x} \times \boldsymbol{J}_B. \tag{4.2.28}$$

These are generally written in the following form;

$$\boldsymbol{\mu}_{I=1} = \frac{g_{I=1}}{2M_N}\frac{\boldsymbol{\sigma}}{2} \otimes \tau_3 \tag{4.2.29}$$

$$\boldsymbol{\mu}_{I=0} = \frac{g_{I=0}}{2M_N}\frac{\boldsymbol{\sigma}}{2}, \tag{4.2.30}$$

where $g_{I=1,0}$ is the isovector and isoscalar g-factor, respectively, and M_N is the nucleon mass. The magnetic moment of a proton or neutron is also defined as

$$\boldsymbol{\mu}_p = \frac{g_p}{2M_N}\frac{\boldsymbol{\sigma}}{2} = \frac{1}{2}\frac{g_{I=1} + g_{I=0}}{2M_N}\frac{\boldsymbol{\sigma}}{2} \tag{4.2.31}$$

$$\boldsymbol{\mu}_n = \frac{g_p}{2M_N}\frac{\boldsymbol{\sigma}}{2} = \frac{1}{2}\frac{g_{I=1} + g_{I=0}}{2M_N}\frac{\boldsymbol{\sigma}}{2}, \tag{4.2.32}$$

where $g_{p,n}$ is the g-factor of the proton and neutron.

■ GKP-Witten current

The isovector magnetic moment was calculated to be

$$\mu^i_{I=1} = \epsilon^{ijk}\int d^3x x^j \text{tr}(J^k_V \tau^3) = -4\pi^2\kappa\rho^2 \text{tr}(W\tau^i W^{-1}\tau^3), \tag{4.2.33}$$

where we performed the same procedure as when we calculated the baryon number charge (4.2.5). If we use the relation,

$$\begin{aligned}\left\langle N'|\mathrm{tr}(t_i W^{-1} t_a W)|N\right\rangle &= \frac{1}{4}\left\langle N'|\mathrm{tr}(\sigma_i W^{-1}\tau_a W)|N\right\rangle \\ &= -\frac{1}{4}\frac{2}{3}\left\langle N'|\sigma_i \otimes \tau_a|N\right\rangle,\end{aligned} \tag{4.2.34}$$

valid for nucleon states (Appendix 4.4.24), $\mu^i_{I=!}$ is calculated to be

$$\frac{16\pi^2\kappa}{3}\rho^2\frac{\boldsymbol{\sigma}}{2}\otimes\tau_3. \tag{4.2.35}$$

Therefore, the g-factor is obtain

$$\begin{aligned}g_{I=1} &= 2M_N\frac{16\pi^2\kappa}{3}\left\langle\rho^2\right\rangle = 2M_N\frac{16\pi^2\kappa}{3}\frac{\sqrt{5}+2\sqrt{5+N_c^2}}{2N_c}\rho_{\mathrm{cl}}^2 \\ &= 2M_N\frac{16\pi^2\kappa}{3}\frac{\sqrt{5}+2\sqrt{5+N_c^2}}{2N_c}\frac{N_c}{8\pi^2\kappa}\sqrt{\frac{6}{5}} \\ &= \frac{4M_N}{M_K K}\frac{1+2\sqrt{1+N_c^2/5}}{6} \simeq 7.03,\end{aligned} \tag{4.2.36}$$

where M_{KK} dependence was recovered by dimensional analysis in the last expression.

Next, we calculate the isoscalar magnetic moment. It is easy that only the last term in (4.1.80) contributes to the following integral:

$$\begin{aligned}\mu^i_{I=0} &= \frac{1}{2}\epsilon^{ijk}\int d^3x x^j J^k_B = \frac{1}{N_c}\epsilon^{ijk}\int d^3x x^j \hat{J}^k_V \\ &= -\frac{1}{N_c}\epsilon^{ijk}\int d^3x x^j \frac{N_c}{2}\frac{\rho^2\chi^a}{4}\epsilon^{kal}\partial_j G^V \\ &= -\frac{\rho^2\chi^i}{12}\int_0^\infty dr 4\pi r^3\partial_r J^0_B(r) = \frac{\rho^2\chi^i}{4} = \frac{J^i}{16\pi^2\kappa} \\ &= \frac{\sigma^i}{32\pi^2\kappa}\end{aligned} \tag{4.2.37}$$

From the above, g-factor is

$$g_{I=0} = \frac{M_N}{8\pi^2\kappa} \simeq 1.68. \tag{4.2.38}$$

The experimental value is $g^{\mathrm{exp}}_{I=0} \simeq 1.76$, which gives a quite good prediction.

■ The 5d Noether current

The isovector magnetic moment can be calculated as

$$\mu^i_{I=1} = \epsilon^{ijk} \int d^3x x^j j^k_{V,a=3}(t, x), \tag{4.2.39}$$

First, we need the current $j^i_{V,a}(t, x)$. In the present case, we ignore the contributions coming from the warp factor and CS term, since they are $1/\lambda$ subleading. Therefore, the only current to consider now is

$$J^i_{V,a}(x, z) = -2\kappa \mathrm{tr}\Big[\Big(\sum_{M=j,z} i[F^{\mathrm{cl}}_{iM}, A^{\mathrm{cl}}_M]\Big) W^{-1} t_a W\Big]. \tag{4.2.40}$$

Here, using

$$\sum_{M=j,z} i[F^{\mathrm{cl}}_{iM}, A^{\mathrm{cl}}_M] = \frac{16\rho^4}{\xi^2(\xi^2+\rho^2)^3}(zt_i - \epsilon_{ija}x^j t_a), \tag{4.2.41}$$

the current is calculated to be

$$\begin{aligned} j^i_{V,a}(t, x) &= \int_{-\infty}^{\infty} dz J^i_{V,a}(x, z) \\ &= \frac{4\pi\kappa}{\rho^2}\Big(\frac{8}{r} - \frac{8r^4 + 20\rho^2 r^2 + 15\rho^4}{(r^2+\rho^2)^{5/2}}\Big)\epsilon_{ijk}x^j \mathrm{tr}(t_k W^{-1} t_a W). \end{aligned} \tag{4.2.42}$$

Using this, the isovector magnetic moment is calculated to be

$$\mu^i_{I=1} = \epsilon_{ijk} \int d^3x x^j j^k_{V,a=3}(t, x) = -8\pi^2\kappa\rho^2 \mathrm{tr}\big(t_i W(t)^{-1} t_3 W(t)\big), \tag{4.2.43}$$

then we obtain the g-factor

$$g_{I=1} = \frac{16\pi^2\kappa}{3} M_N \langle\rho^2\rangle = \sqrt{\frac{2}{3}}\Big(1 + 2\sqrt{1 + 2\sqrt{1 + \frac{N_c^2}{5}}}\Big) M_N = 6.83. \tag{4.2.44}$$

This result is inconsistent with the experimental data and closer to the Skyrme model results than the (4.2.36) results.

Next, the isoscalar magnetic moment is calculated as

$$\mu^i_{I=0} = \frac{1}{2}\epsilon_{ijk} \int d^3x x^j j^k_B(t, x). \tag{4.2.45}$$

Again, using the current

$$\begin{aligned} J_B^i(x,z) &= -\frac{1}{8\pi^2}\epsilon_{ijk}\mathrm{tr}(F_{jk}F_{0z}+2F_{0j}F_{kz}) \\ &= \frac{3}{\pi^2}\frac{\rho^4}{(\xi^2+\rho^2)^4}\left[(\delta_{ia}z-\epsilon_{ija}x^j)\chi^a(t)+2x^i\frac{d}{dt}\ln\rho(t)\right] \end{aligned} \tag{4.2.46}$$

ignoring the contributions from the warp factor and CS term, the subleading terms of $1/\lambda$, the 4-dimensional current is given as

$$\begin{aligned} j_B^i(t,x) &= \int_{-\infty}^{\infty} dz J_B^i(x,z) \\ &= \frac{15}{16\pi}\frac{\rho^2}{(r^2+\rho^2)^{7/2}}\Big(-\epsilon_{ijk}x^j\chi^a(t)+2x^i\frac{d}{dt}\ln\rho(t)\Big). \end{aligned} \tag{4.2.47}$$

From this, the isoscalar magnetic moment is calculated to be

$$\mu_{I=0}^i = \frac{1}{2}\epsilon_{ijk}\int d^3x x^j j_B^k(t,x) = \frac{\rho^2}{4}\chi^i(t) = \frac{1}{16\pi^2\kappa}J_i, \tag{4.2.48}$$

so the g-factor is given as

$$g_{I=0} = \frac{M_N}{8\pi^2\kappa} = 1. \tag{4.2.49}$$

Again, this value is much smaller than the prediction obtained from the GKP-Witten current and does not agree with the experimental data.

4.2.4 Axial Coupling and Axial Radius

In the non-relativistic $\boldsymbol{k}\to 0$ limit, the axial coupling of the nucleon is defined by using the axial current $J_A^i(x)$ as follows;

$$\begin{aligned} &\int d^3x e^{i\boldsymbol{k}\cdot\boldsymbol{x}}\left\langle N,s_3'I_3'|J_A^{ai}|N,s_3,I_3\right\rangle\times 2 \\ &= \frac{2}{3}g_A^{NN}(\boldsymbol{k}=0)\left(\sigma^a\right)_{s_3',s_3}\left(\tau^a\right)_{I_3',I_3}, \end{aligned} \tag{4.2.50}$$

where the factor 2/3 on the right-hand side is needed in the chiral limit [14]. Here, axial coupling is related to coupling $g_{\pi NN}$ of πNN through the Goldberger-Treiman relation

$$g_A^{NN} = \frac{f_\pi g_{\pi NN}}{M_N}. \tag{4.2.51}$$

■ GKP-Witten current

With axial current (4.1.82), the matrix elements of axial current are computed as

$$\int d^3x J_A^{a,i} = \frac{4}{3}\pi^2\kappa\rho^2 \mathrm{tr}(W\tau^i W^{-1}\tau^a) \int d^3x \partial_j^2 H^A. \tag{4.2.52}$$

Using (4.1.58), $\partial_j^2 H^A$ transformed to

$$\begin{aligned} \partial_j^2 H^A &= \partial_j^2\Big(\frac{1}{2\pi^2}\frac{1}{k(Z)}\frac{1}{r} - \sum_{n=1}^{\infty}\frac{g_{a^n}}{\lambda_{2n}}\partial_Z\psi_{2n}(Z)Y_{2n}(r)\Big) \\ &= \Big(4\pi\frac{1}{2\pi^2}\frac{1}{k(Z)} - \sum_{n=1}^{\infty}\frac{g_{a^n}}{\lambda_{2n}}\partial_Z\psi_{2n}(Z)\Big)\delta^3(\vec{x}-\vec{X}) - \sum_{n=1}^{\infty} g_{a^n}\partial_Z\psi_{2n}(Z)Y_{2n}(r), \end{aligned} \tag{4.2.53}$$

the first and second terms of (4.2.53) cancel each other because

$$\begin{aligned} &\sum_{n=1}^{\infty}\frac{g_{a^n}}{\lambda_{2n}}\langle\partial_Z\psi_{2n}(Z)\rangle = \partial_Z\int dz\psi_0(z)\kappa h(z)\sum_{n=1}^{\infty}\psi_{2n}(z)\langle\psi_{2n}(Z)\rangle \\ &= \partial_Z\int dz\psi_0(z)\delta(z-Z) = \partial_Z\psi_0(Z) = \frac{2}{\pi}\frac{1}{k(Z)} \end{aligned} \tag{4.2.54}$$

is satisfy by the same argument as (4.2.5). Because of

$$\begin{aligned} \int d^3x\partial_j^2\langle H^A\rangle &= -\sum_{n=1}^{\infty} g_{a^n}\langle\partial_Z\psi_{2n}(Z)\rangle\int_0^{\infty} dr 4\pi r^2\Big(-\frac{1}{4\pi}\frac{e^{-\sqrt{\lambda_{2n}}r}}{r}\Big) \\ &= \sum_{n=1}^{\infty} g_{a^n}\partial_Z\psi_{2n}(Z)\int_0^{\infty} drre^{-\sqrt{\lambda_n}r} = \frac{2}{\pi}\frac{1}{k(Z)}, \end{aligned} \tag{4.2.55}$$

g_A^{NN} is obtained by

$$g_A^{NN} = \frac{16\pi\kappa}{3}\Big\langle\frac{\rho^2}{k(Z)}\Big\rangle, \tag{4.2.56}$$

where (4.2.5) is used. If a numerical calculation is performed, the axial coupling is computed as

$$g_A^{NN} \simeq 0.734, \tag{4.2.57}$$

where the experimental value is $g_A^{NN}|_{\mathrm{exp}} \simeq 1.27$, indicating that the predicted value of this model is considerably smaller than the experimental value.

■ The 5d Noether current

In leading of $1/\lambda$ expansion, the effect of the warp factor and the contribution from the CS term can be ignored, hence the 5d axial current is calculated to be

$$J^i_{A,a}(x,z) = -\frac{8\kappa\rho^2}{\xi^2(\xi^2+\rho^2)^2}\left[\frac{4\rho^2 z\psi_A(z)}{\xi^2+\rho^2} + \left(z^2 - \frac{x^2}{3}\frac{\psi_A(z)}{dz}\right)\right]\mathrm{tr}\left(t_i W(t)^{-1} t_a W(t)\right). \tag{4.2.58}$$

Although $\psi_A(z)$ is written as $\psi_A(z) = (2/\pi)\arctan z$, as far as leading of the $1/\lambda$ expansion is concerned, we can treat $\psi_A(z) \simeq (2/\pi)z$ as well as the effect of the warp factor. Thus, the 4-dimensional current can be written as

$$\begin{aligned} j^i_{A,a}(t,x) &= \int_{-\infty}^{\infty} dz J^i_{A,a}(x,z) \\ &= \frac{32\kappa}{3\rho^2}\left[8r - \frac{8r^4+12\rho^2r^2+3\rho^4}{(r^2+\rho^2)^{3/2}}\right]\mathrm{tr}\left(t_i W(t)^{-1} t_a W\right). \end{aligned} \tag{4.2.59}$$

Since this 4-dimensional current becomes

$$\int d^3x j^i_{A,a}(x,t) = -\frac{32\pi\kappa\rho^2}{3}\left(t_i W(t)^{-1} t_a W\right), \tag{4.2.60}$$

by spatial integration, axial coupling is calculated as

$$g_A = \frac{16\pi\kappa}{3}\left\langle\rho^2\right\rangle. \tag{4.2.61}$$

From this, we obtain the axial coupling

$$g_A = \frac{\sqrt{6}}{3\pi}\left(1 + 2\sqrt{1+\frac{N_c^2}{5}}\right) = 1.13. \tag{4.2.62}$$

Remarkably, in the leading calculation of the $1/\lambda$ expansion, this value does not depend on the parameters of the model. Nevertheless, we find that it predicts a value quite close to the experimental data. Finally, we summarize in Table 4.1 the values of the physical quantities for the static properties of the nucleon obtained from the two currents and their experimental values, together with the results of the Skyrme model.

4.3 Dynamical Properties of the Nucleon Resonances

In the previous section, we reviewed static properties of nucleons, which have been carried out by [1, 2], etc. However, to understand the nature of the baryon resonances,

Table 4.1 Results from the each analysis on the static properties of the nucleon

	GKP-Witten current	5d Noether current	ANW [8]	Exp
$\sqrt{\langle r^2\rangle_{I=0}}$ (fm)	0.785	0.82	0.59	0.806
$\sqrt{\langle r^2\rangle_{E,p}}$ (fm)	0.785	–	–	0.875
$\sqrt{\langle r^2\rangle_{E,n}}$ (fm)	0	–	–	−0.116
$g_{I=1}$	7.03	6.83	6.38	9.41
$g_{I=0}$	1.68	1	1.11	1.76
g_A^{NN}	0.734	1.13	0.61	1.27

it is also necessary to reveal their dynamical properties. When considering dynamical processes of baryons, it is important to include interactions involving pions, because many baryon resonances are formed and decay through pions. The contribution of mesons other than pions is also important, and it has been shown that many phenomena can be explained by a hybrid structure where constituent quarks form the core structure which is dressed by meson clouds [9, 10].

The Sakai-Sugimoto model leads to an effective model of hadrons as a flavor gauge theory in five dimensions as a holographic dual of massless QCD. The gauge field, which plays the fundamental role in this model, originated from an open string with both ends in the D8 brane carrying the flavor, and by mode expansion, we identify each mode, including pions, with an infinite number of vector/axial-vector mesons. According to Chap. 2, baryons are interpreted as instantons on the D8 brane [11–13]. The dynamics of baryons at low energies are given by collective motions of instantons/solitons, which is quite different from the quark model based on a single particle picture of quarks. Moreover, in the low energy limit, it is known that the Skyrmion [14] is derived by projecting an instanton in the five-dimensional spacetime onto 4-dimensional spacetime using the Atiyah-Manton method [7]. As mentioned in Chap. 2, these baryon pictures of the Sakai-Sugimoto model are closely related to the meson cloud picture, which has been found in the study of nucleon resonances so far. A particularly remarkable fact in this model is that the masses of the Roper resonance and the negative parity state degenerate in the resulting mass formula. This shows that it captures the features of the hadron spectroscopy better than the other models. We consider it a worthwhile attempt to investigate the dynamical properties of nucleon resonance using the Sakai-Sugimoto model, which has the above features. This attempt is also expected to be a milestone in the development of the study of dynamical properties of nucleon resonance using holographic QCD. In this section, in Sect. 4.3.1, we first calculate the axial coupling and the decay width of one pion emission calculated from this coupling using GKP-Witten current. Next, in Sect. 4.3.2, we attempt to analyze the electromagnetic form factor. Finally, in Sect. 4.3.3, we summarize the results.

4.3.1 The Axial Coupling and One Pion Emission

In this subsection, we investigate the one-pion emission processes of nucleon resonances using the GKP-Witten current of the Sakai-Sugimoto model. In particular, we aim to calculate the axial coupling and decay width.

4.3.1.1 Formulation

By evaluating the matrix elements of the axial vector current, the axial coupling $g_A^{NN^*}$ is obtained. The axial coupling is related to the coupling constant $g_{\pi NN^*}$ of πNN^* through the Goldberger-Treiman relation, one of the low energy theorems. From this $g_{\pi NN^*}$, the decay width of one pion emission is calculated.

It turns out that the equation of axial coupling is completely different for positive and negative parity resonance states. In the non-relativistic limit, the axial current between the nucleon N and the positive/negative parity excited state $N^*_{\text{even/odd}}$ is defined by

$$J_A^\mu = \overline{\psi}_N \gamma^\mu \psi_{N^*_{\text{even}}}, \tag{4.3.1}$$

$$J_A^\mu = \overline{\psi}_N \gamma^\mu \gamma_5 \psi_{N^*_{\text{odd}}}, \tag{4.3.2}$$

with

$$\psi_B = \begin{pmatrix} u_B \\ 0 \end{pmatrix}, \quad B = N, N^*_{\text{even}}, N^*_{\text{odd}}, \tag{4.3.3}$$

where γ_5 is required if the parity of the initial and final states are the same, but not if they are different. From the above, in the case of $B = N^*_{\text{even}}$, only the spatial component of the current, and in the case of N^*_{odd}, only the time component of the current contributes to the current in the relativistic limit.

Therefore, in the following, we present the formulation of the axial coupling and the decay width in each case.

■ Positive parity N^*_{even}

The axial coupling $g_A^{NN^*_{\text{even}}}(\boldsymbol{k})$ for the transition $N^*_{\text{even}} \to N + \pi$ is defined as follows:

$$\int d^3x e^{i\boldsymbol{k}\cdot\boldsymbol{x}} \langle N, s_3' I_3' | J_A^{ai} | N^*_{\text{even}}, s_3, I_3 \rangle \times 2$$
$$= g_A^{NN^*_{\text{even}}}(\boldsymbol{k}) \left(\delta_{ia} - \frac{k_i k_a}{\boldsymbol{k}^2} \right) (\sigma^a)_{s_3', s_3} (\tau^a)_{I_3', I_3}. \tag{4.3.4}$$

in the chiral limit, this is written as

$$
\begin{aligned}
&\int d^3x \left\langle N, s_3', I_3' \right| J_A^{ai} \left| N^*_{\text{even}}, s_3, I_3 \right\rangle \times 2 \\
&= \frac{2}{3} g_A^{NN^*_{\text{even}}} \left(\sigma^a\right)_{s_3', s_3} \left(\tau^a\right)_{I_3', I_3},
\end{aligned} \tag{4.3.5}
$$

where the factor 2/3 on the right-hand side is needed [8].

The decay width of $N^*_{\text{even}} \to N + \pi$ can be computed by the formula

$$
\begin{aligned}
\Gamma_{N^*_{\text{even}} \to N+\pi} =& \frac{1}{2M_{N^*_{\text{even}}}} \int \frac{d^3 p_N}{(2\pi)^3\, 2E_N} \frac{d^3 p_\pi}{(2\pi)^3\, 2E_\pi} \\
&\times (2\pi)^4\, \delta^4 \left(p_N + p_\pi\right) \left|t_{fi}\right|^2,
\end{aligned} \tag{4.3.6}
$$

where the amplitude t_{fi} is given by the Lagrangian

$$
L = i \frac{M_N + M_{N^*}}{2 f_\pi} g_A^{NN^*_{\text{even}}} \overline{\psi}_{N^*_{\text{even}}} \gamma_5 \vec{\tau} \cdot \vec{\pi} \psi_N + h.c., \tag{4.3.7}
$$

as follows

$$
\begin{aligned}
t_{fi} =& \left\langle N(-\vec{k}) \pi(\vec{k}) \right| L \left| N^*_{\text{even}}(\vec{0}) \right\rangle \\
=& \sqrt{2M_{N^*_{\text{even}}}} \sqrt{M_N + E_N} \\
&\times \frac{M_N + M_{N^*_{\text{even}}}}{2 f_\pi} \frac{g_A^{NN^*_{\text{even}}}}{E_N + M_N} \left\langle s_3' \right| \vec{\sigma} \cdot \vec{k} \left| s_3 \right\rangle.
\end{aligned} \tag{4.3.8}
$$

Here we have expressed the effective $\pi N N^*_{\text{even}}$ coupling $g_{\pi N N^*_{\text{even}}}$ in terms of the axial coupling by using the Goldberger-Treiman relation,

$$
g_A^{NN^*_{\text{even}}} = \frac{f_\pi g_{\pi N N^*_{\text{even}}}}{\left(M_N + M_{N^*_{\text{even}}}\right)/2}. \tag{4.3.9}
$$

Hence we obtain

$$
\begin{aligned}
&\Gamma_{N^*_{\text{even}} \to N+\pi} \\
&= \frac{k}{4\pi} \frac{M_N + E_N}{M_{N^*_{\text{even}}}} \left(\frac{M_N + M_{N^*_{\text{even}}}}{2 f_\pi} \frac{g_A^{NN^*_{\text{even}}} k}{E_N + M_N} \right)^2.
\end{aligned} \tag{4.3.10}
$$

where k is the momentum of the decaying pion and is given by

$$
k = \frac{\left(M^2_{N^*_{\text{even}}} - (M_N + m_\pi)^2\right)^{1/2} \left(M^2_{N^*_{\text{even}}} - (M_N - m_\pi)^2\right)^{1/2}}{2M_{N^*_{\text{even}}}} \tag{4.3.11}
$$

■ Negative parity transition

The axial coupling $g_A^{NN^*_{\rm odd}}(\boldsymbol{k})$ for the transition $N^*_{\rm odd} \to N + \pi$ is defined as follows:

$$g_A^{NN^*_{\rm odd}}(\tau^a)_{I_3 I_3'} = \int d^3x e^{i\boldsymbol{k}\cdot\boldsymbol{x}} \left\langle N, I_3' | j_A^{a0} | N^*_{\rm odd}, I_3 \right\rangle \times 2 \tag{4.3.12}$$

The decay width of $N^*_{\rm odd} \to N + \pi$ can be computed by the formula

$$\Gamma_{N^*_{\rm odd} \to N\pi} = \frac{1}{2M_{N^*_{\rm odd}}} \int \frac{d^3p_N}{(2\pi)^3 2E_N} \frac{d^3p_\pi}{(2\pi)^3 2E_\pi} (2\pi)^4 \delta^4(p_N + p_\pi) |t_{fi}|^2 \tag{4.3.13}$$

The transition matrix t_{fi} of the Lagrangian

$$\mathcal{L} = i \frac{g_A^{NN^*_{\rm odd}}}{2f_\pi} \overline{\psi}_N \gamma^0 \partial_0 \pi^a \tau^a \psi_{N^*_{\rm odd}}, \tag{4.3.14}$$

is defined as

$$\begin{aligned} t_{fi} &= \left\langle N(-k); \pi(k) | \mathcal{L} | N^*_{\rm odd}(0) \right\rangle \\ &= i \frac{g_A^{NN^*_{\rm odd}}}{2f_\pi} E_\pi \sqrt{E_N + M_N} \sqrt{2m_{M^*_{\rm odd}}} \delta_{s_3 s_3'}. \end{aligned} \tag{4.3.15}$$

Hence we obtain

$$\Gamma_{N^*_{\rm odd} \to N\pi} = \frac{(g_A^{NN^*_{\rm odd}})^2}{16\pi} \frac{|\boldsymbol{k}| E_\pi^2 (E_\pi + m_N)}{f_\pi^2 M_{N^*_{\rm odd}}} \tag{4.3.16}$$

Here we have expressed the effective $\pi N N^*_{\rm even}$ coupling $g_{\pi N N^*_{\rm even}}$ in terms of the axial coupling by using the Goldberger-Treiman relation,

$$g_A^{NN^*_{\rm odd}} = \frac{f_\pi g_{\pi N N^*_{\rm odd}}}{\left(M_{N^*_{\rm odd}} - M_N\right)/2}. \tag{4.3.17}$$

4.3.1.2 Results

There are two parameters of this model, M_{KK} and κ. Following Ref. [8] they are determined to reproduce the mass of the ρ (776 MeV) meson, and the pion decay constant $f_\pi = 92.4$ MeV,

$$M_{KK} = 940 \text{ MeV}, \quad \kappa = 0.00745 \tag{4.3.18}$$

In Table 4.2, we also summarize the results with the parameter set used in Ref. [2].

Now that we are prepared to calculate the axial coupling and decay width for one pion emission, we will present the results for the low-lying nucleon resonances described in Chap. 1.

■ Roper resonance

First, we obtain the axial coupling g_A for the transition from the Roper resonance to the nucleon. In momentum space, the axial current is given by

$$\tilde{J}^{\mu}_{A}(\vec{q}) = \int d^3x\, e^{-i\vec{q}\cdot\vec{x}} J^{\mu}_{A}(r),$$
$$\tilde{J}^{cj}_{A}(\vec{q}) = e^{-i\vec{q}\cdot\vec{X}} 2\pi^2\kappa\rho^2 \mathrm{tr}\left(\tau^c W\tau^a W^{-1}\right) \times \left(\delta_{aj} - \frac{k_a k_j}{\vec{k}^2}\right)\sum_{n\geq 1}\frac{g_{a^n}\partial_Z\psi_{2n}(Z)}{\vec{k}^2+\lambda_{2n}}, \tag{4.3.19}$$

where the following equation is used,

$$\int d^3x e^{-i\vec{k}\cdot\vec{x}} Y_n(|\vec{x}-\vec{X}|) = -e^{-i\vec{k}\cdot\vec{X}}\frac{1}{\vec{k}^2+\lambda_n}. \tag{4.3.20}$$

$$\int d^3x e^{-i\vec{k}\cdot\vec{x}} H^A(Z,|\vec{x}-\vec{X}|) = -e^{-i\vec{k}\cdot\vec{x}}\frac{1}{\vec{k}^2}\sum_{n=1}^{\infty}\frac{g_{a^n}\partial_Z\psi_{2n}(Z)}{\vec{k}^2+\lambda_{2n}}. \tag{4.3.21}$$

Using (4.3.19), (2.4.121), (2.4.122) and (4.4.24), we obtain

$$g_A^{NN^*(1440)}(\vec{q}) = \frac{8\pi^2\kappa}{3}\left\langle R_{n_\rho=1}|\rho^2|R_{n_\rho=0}\right\rangle\sum_{n=1}\frac{g_{a_n}\langle\partial_Z\psi_{2n}(Z)\rangle}{\vec{k}^2+\lambda_{2n}}. \tag{4.3.22}$$

where $\langle\partial_Z\psi_{2n}(Z)\rangle$ stands for the expectation value using the wave functions of Z and R_{n_ρ} is the confluent hypergeometric functions (2.4.121), which normalized as follow,

$$\int_0^{\infty} d\rho\rho^3 R_{n_\rho}(\rho)^2 = 1. \tag{4.3.23}$$

The matrix element of ρ^2 can be computed and the result is

$$\left\langle R_{N^*(1440)}|\rho^2|R_N\right\rangle = \left(1+2\sqrt{1+\frac{N_c^2}{5}}\right)^{-1/2}\left\langle R_N|\rho^2|R_N\right\rangle = \frac{\sqrt{5}}{2N_c}\left(1+2\sqrt{1+\frac{N_c^2}{5}}\right)^{1/2}\rho_{cl}^2 \tag{4.3.24}$$

with ρ_{cl} being the classical instanton size given by

$$\rho_{cl}^2 = \frac{N_c}{8\pi^2\kappa}\sqrt{\frac{6}{5}}. \tag{4.3.25}$$

We notice that the transition matrix elements of $N^*(1440) \to N + \pi$ are related to the nucleon matrix elements, which is an interesting feature of the present model associated with the collective nature of the baryons. The axial coupling constant is then defined $g_A^{NN^*(1440)} = g_A^{NN^*(1440)}(\vec{0})$ at $\vec{k} = 0$. Using the relation (4.2.5),

$$\sum_{n=1} \frac{g_{a_n}\partial_Z\psi_{2n}(Z)}{\lambda_{2n}} = \frac{2}{\pi}\frac{1}{k(Z)}, \tag{4.3.26}$$

$g_A^{NN^*(1440)}$ can be expressed in a compact form:

$$g_A^{NN^*(1440)} = \frac{16\pi\kappa}{3}\langle R_{N^*}|\rho^2|R_N\rangle\left\langle\frac{1}{k(Z)}\right\rangle. \tag{4.3.27}$$

Using the two parameters mentioned in (4.3.18), the prediction of the present model for $g_A^{NN^*(1440)}$ is

$$g_A^{NN^*(1440)} = 0.352. \tag{4.3.28}$$

With the axial coupling (4.3.28), we investigate the one pion emission decay $N^*(1440) \to \pi N$. Before we go on, we will give an explanation of the experimental values. Because the Roper resonance has a very large width causing uncertainties in the Breit-Wigner fitting, we refer to the result of the pole analysis. Following the PDG table [15], we quote the following nominal values

$$\begin{aligned} M_{N^*(1440)} &= 1360 - 1380\ (\sim 1370)\ \text{MeV}, \\ \Gamma_{\text{total}} &= 160 - 190\ (\sim 175)\ \text{MeV}, \end{aligned} \tag{4.3.29}$$

and the branching ratio of the one pion decay

$$N^*(1440) \to N\pi:\ \ 55 - 75\ \%. \tag{4.3.30}$$

Using the lower and upper bounds for the total decay width and branching ratio, we find the partial decay width of the one pion decay

$$\Gamma_{N^*(1440)\to\pi N} \sim 90 - 140\ \text{MeV}. \tag{4.3.31}$$

Using $M_N = 940$ MeV, $M_{N^*} = 1370$ MeV, $m_\pi = 140$ MeV (pion mass), $k = 342$ MeV, we find

$$\Gamma_{N^*(1440)\to N+\pi} = 49 \text{ MeV} \tag{4.3.32}$$

By considering the form factor effect, the $g_A^{NN^*(1440)}$ value at $\vec{k} = 342$ MeV becomes about 13 % smaller, and hence $\Gamma_{N^*(1440)\to N+\pi} \sim 43$ MeV (The form factor is given in Fig. 1 in Ref. [1].). If we use $M_{N^*} = 1440$ MeV and $k = 398$ MeV [15], we find 77 MeV for (4.3.43) and 64 MeV for the finite k. These estimations show that there is ambiguity in comparison with actual experimental data due to uncertainties in the exact resonance point.

Our predictions of the decay width obtained from different resonance points is are smaller than the experimental value (4.3.31). This is because the nucleon axial coupling g_A^{NN} is small. In fact, the nucleon g_A^{NN} is computed in a similar manner as for $g_A^{NN^*(1440)}$. The result is

$$g_A^{NN} = 0.732. \tag{4.3.33}$$

This value is significantly smaller than the experimental value $g_A^{NN} = 1.25$. The small g_A is a common problem of the solitonic description of baryons. One possible resolution to recover the experimental value $g_A^{NN} = 1.25$ is to take into account $1/N_c$ corrections (Ref. [16] and references there). Here, we will not discuss this point further and point out some interesting features of the present model. We find that the following relation holds between the axial coupling of the nucleon and that of the transition from Roper to the nucleon.

$$\begin{aligned} g_A^{NN^*(1440)}/g_A^{NN} &= \left(1 + 2\sqrt{1 + \frac{N_c^2}{5}}\right)^{-1/2} \\ &= 0.5. \end{aligned} \tag{4.3.34}$$

It is worth remarking here that this relation is independent of the parameters of the model. If we determine the value of $g_A^{NN^*(1440)}$ to derive the experimental value of $\Gamma_{N^*\to\pi N} \sim 110$ MeV, we find the ratio

$$g_A^{NN^*(1440)}/g_A^{NN} = 0.77/1.25 \sim 0.6, \tag{4.3.35}$$

which agrees well with the present model prediction within ~ 20 % accuracy, whose agreement is better than the absolute value.

■ Negative parity resonance $N^*(1535)$

It is obvious from the structure of the axial current that the spatial component of the axial current does not contribute to the axial coupling $g_A^{NN^*_{\text{odd}}}$ due to the parity determined by the wave function of Z and the completely anti-symmetric tensor ϵ_{ibj}. Therefore, only the time component of the axial current contributes in the negative parity transition in the non-relativistic limit [17]. The time component of the axial current is written as

$$\tilde{J}^0_{A,a}(\boldsymbol{k}) = 2\pi^2\kappa\Big[i\frac{k_a}{\boldsymbol{k}^2}\mathrm{tr}[\tau^a\partial_0(\rho^2 W\tau^b W^{-1})]\Big]e^{-ik\cdot X}\sum_{n=1}^{\infty}\frac{g_{a^n}\partial_Z\psi_{2n}}{\boldsymbol{k}^2+\lambda_{2n}}$$
$$+\Big(I_a - 2\pi^2\kappa\rho^2\frac{P_i}{M_0}\epsilon^{iaj}k_j\Big)e^{-ik\cdot X}\sum_{n=1}^{\infty}\frac{g_{a^n}\psi_{2n}}{\boldsymbol{k}^2+\lambda_{2n}}, \tag{4.3.36}$$
$$\tilde{J}^i_{A,a}(\boldsymbol{k}) = 2\pi^2\kappa\rho^2\mathrm{tr}(\tau^a W\tau^b W^{-1})e^{-ik\cdot X}\Big[\Big(\delta_{bi} - \frac{k_b k_i}{\boldsymbol{k}^2}\Big)\sum_{n=1}^{\infty}\frac{g_{a^n}\partial_Z\psi_{2n}}{\boldsymbol{k}^2+\lambda_{2n}}$$
$$-i\epsilon^{ibj}k_j\sum_{n=1}^{\infty}\frac{g_{a^n}\psi_{2n}}{\boldsymbol{k}^2+\lambda_{2n}}\Big], \tag{4.3.37}$$

in the momentum space. The axial coupling at the transition momentum $\vec{k}=0$ is expressed as

$$g_A^{NN^*(1535)}(\boldsymbol{k}^2) = \int_{-\infty}^{\infty} dZ\psi_{n_Z=0}(Z)\psi_{n_Z=1}(Z)\sum_{n=1}^{\infty}\frac{g_{a^n}\psi_{2n}}{\boldsymbol{k}^2+\lambda_{2n}}, \tag{4.3.38}$$

using this axial current and the wave functions (2.4.123) and (2.4.124) which are normalized as follows,

$$\int_{-\infty}^{\infty} dZ\psi_{n_Z}(Z)^2 = 1. \tag{4.3.39}$$

The axial coupling constant is defined by $\vec{k}=0$ and $g_A^{NN^*(1535)} = g_A^{NN^*(1535)}(\vec{0})$. Using the relation (4.2.54), we obtain

$$g_A^{NN^*(1535)} = \int_{-\infty}^{\infty} dZ\psi_{n_Z=1}(Z)\psi_{n_Z=0}(Z)\psi_0(Z). \tag{4.3.40}$$

We analytically obtain the following result;

$$g_A^{NN^*(1535)} = \sqrt{\frac{2}{\pi}}e^{2M_0/\sqrt{6}}\mathrm{erfc}\Big(\sqrt{\frac{2}{\sqrt{6}}M_0}\Big), \tag{4.3.41}$$

where $\mathrm{erfc}(x)$ is a complementary error function. From the above, we obtain the following values as axial coupling,

$$g_A^{NN^*(1535)} = 0.42. \tag{4.3.42}$$

Using $M_N = 940$ MeV, $M_{N^*} = 1510$ MeV, $m_\pi = 140$ MeV (pion mass), $q = 448$ MeV, we find

$$\Gamma_{N^*(1535)\to N+\pi} = 54\ \mathrm{MeV}, \tag{4.3.43}$$

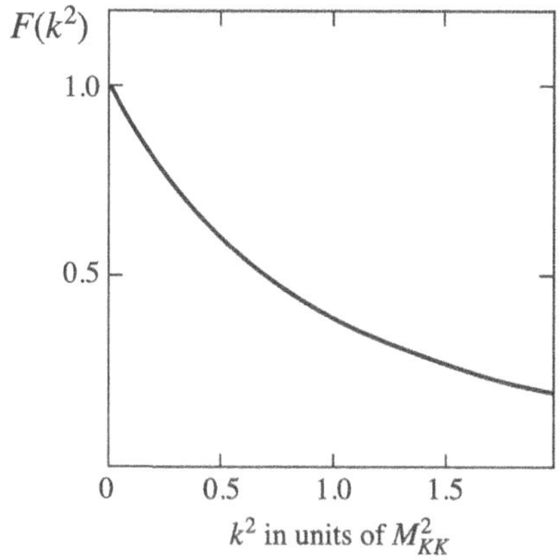

Fig. 4.1 The emitted pion momentum $\boldsymbol{k}$ dependence of the form factor (Reprinted from [17], by the Authors licensed under Creative Commons Attribution 4.0 International license)

where the mass of the resonant state is the result from pole position. In this computation the value of $g_A^{NN^*}$ at $\vec{q} = 0$ is used. By considering the form factor effect, the $g_A^{NN^*}$ value at $\vec{q} = 448$ MeV becomes about 25 % smaller, and hence $\Gamma_{N^*(1535)\to N+\pi} \sim 30$ MeV (Fig. 4.1). According to PDG, the total decay width of the negative parity resonance $N^*(1535)$ is 130 MeV, and the branching ratio of the one pion emission is $32 \sim 52\%$, so the partial decay width is

$$\Gamma^{\exp}_{N^*(1535)\to N\pi} = 42 - 68\text{MeV}. \tag{4.3.44}$$

Although our results are somewhat smaller than the experimental values, This value is in good agreement with the experimental value at the level of accuracy that is to be expected for models of this type.

■ $\Delta(1232)$

Using (4.3.19) and (2.4.121), we obtain

$$g_A^{NN^*}(\vec{q}) = \frac{8\sqrt{2}\pi^2\kappa}{3}\left\langle R_N|\rho^2|R_N\right\rangle \sum_{n=1} \frac{g_{a_n}\,\langle \partial_Z \psi_{2n}(Z)\rangle}{\vec{k}^2 + \lambda_{2n}}, \tag{4.3.45}$$

where we used the following instead of (4.2.34),

$$\left\langle N, I_3', s_3'|\mathrm{tr}(W\tau^i W^{-1}\tau^a)|\Delta^+(1232), I_3 = 1/2, s_3 = 1/2\right\rangle = \frac{2\sqrt{2}}{3}(\sigma^i)_{s_3',1/2}(\tau^a)_{I_3',1/2}$$

(refer to (4.4.25) Appendix 4.4) (4.3.46)

From the above, the axial coupling $g_A^{N\Delta}$ is obtained as

$$g_A^{N\Delta} = 1.18. \tag{4.3.47}$$

Using $M_N = 940$ MeV, $M_{N^*} = 1232$ MeV, $m_\pi = 140$ MeV (pion mass), $k = 229$ MeV, we find

$$\Gamma_{\Delta^+ \to N+\pi} = 160 \text{ MeV} \tag{4.3.48}$$

By considering the form factor effect, the $g_A^{N\Delta}$ value at $\vec{k} = 229$ MeV becomes about 10 % smaller, and hence $\Gamma_{\Delta \to N+\pi} \sim 140$ MeV. Therefore, our prediction reproduces the experimental value of $\Gamma_{\Delta \to N+\pi} \sim 131$ MeV quite well.

4.3.1.3 Discussion

The parameter set used so far was chosen to reproduce the ρ meson mass and the pion decay constant f_π. This parameter set does not reproduce the mass splitting of the baryon resonance. On the other hand, we can reproduce the mass splitting of a nucleon and $\Delta(1232)$ by choosing $M_{KK} = 488$ MeV as in Sect. 2.4. Also we determine $\kappa = 0.0137$ to reproduce the pion decay constant $f_\pi = 64.5$ MeV, following Adkins et al. [8]. In addition to the parameter set in the main discussion, this parameter set is also included and the results are summarized in Table 4.2.

By choosing Set2, a large decay width can be obtained. However, in this case, it can be seen that the form factor decreases rapidly with increasing $\vec{k}$. Therefore, when the effect of the form factor is taken into account, the decay width becomes smaller than the result of Set1.

Thus, the behavior of the result depends on the choice of parameters. Nevertheless, we find that our results reproduce relatively well the decay width of the one-pion emission of the Roper resonance, which is almost zero in the quark model. Furthermore, it should be emphasized that one of the unique and remarkable features of this model is that the relation between the nucleon axial coupling constant g_A^{NN} and $g_A^{NN^*(1440)}$ is independent of the parameters of the model.

So far transitions of one pion emissions have been described well. Turning to the electromagnetic transitions, however, we have found that the GKP-Witten current does not reproduce the experimental data of the Roper resonance well. Therefore, in the next chapter, we will attempt to calculate the electromagnetic transition form factor using the 5d Noether current instead of the GKP-Witten current.

Table 4.2 The axial coupling obtained from each set of parameters

	M_{KK}	κ	g_A^{NN}	$g_A^{NN^*(1440)}$	$g_A^{NN^*(1535)}$	$g_A^{N\Delta(1232)}$
Set1	940 MeV	0.00745	0.732	0.352	0.42	1.18
Set2	488 MeV	0.0137	0.837	0.402	0.35	1.35
EXP.	–	–	1.25	0.77	0.51	1.07

4.3.2 *The Electromagnetic Transition Form Factor*

The pion loop gives an important contribution to the electromagnetic transition amplitude at relatively small momentum transfers because the spontaneous breaking of the chiral symmetry produces Goldstone bosons (pions) that are massless at the chiral limit. The importance of this contribution was already known long ago from the fact that the electromagnetic transition amplitude of $\Delta(1232)$ is underestimated when only the constituent quarks is taken into account [18–20]. Furthermore, recent experiments at the Thomas Jefferson National Accelerator Facility (JLab) have extracted helicity amplitudes wide range of momentum transfer of electromagnetic transitions from the nucleon to the Roper resonance [21–25], which is in serious conflict with the prediction of the naive quark model. To solve these problems, many theoretical efforts have been made [9, 10, 20, 26–28]. It was pointed out that relativistic effects of the confined quarks at a short distance and meson cloud effects at a long distance are important to improve the above-mentioned problems [9, 10]. The Sakai-Sugimoto model justifies the description of baryons as solitons of mesons, and the picture here is like the meson cloud picture itself. Therefore, by analyzing the electromagnetic transition amplitude in this model, the importance of the meson cloud effect can be understood from a different perspective, in terms of the soliton picture.

In particular, the electromagnetic transition amplitude of $A_{1/2}$ of the Roper resonance shows a unique feature not found in other nucleon resonances. The experimental data of $A_{1/2}$ of the Roper resonance is shown in Fig. 4.2. It takes a finite negative value near the real photon point $Q^2 \sim 0$. This feature essentially cannot be explained from the quark model, which has successfully explained the properties of many baryon resonances, as explained in Sect. 4.1. Understanding this behavior is important for revealing low-energy QCD. Therefore, in this subsection, we will analyze it using the 5d Noether current.

4.3.2.1 Formulation

Using the current obtained in Sect. 4.1.3.1, the helicity amplitude is given by

$$A_{3/2}(Q^2) = \sqrt{\frac{2\pi\alpha}{K}} \int d^3x \left\langle \psi_N, s_3 = \frac{3}{2} | \epsilon^{(+)}_\mu j^\mu_{em} | \psi_{N^*}, s_3 = \frac{1}{2} \right\rangle e^{i|\vec{k}|x^3}, \tag{4.3.49}$$

$$A_{1/2}(Q^2) = \sqrt{\frac{2\pi\alpha}{K}} \int d^3x \left\langle \psi_N, s_3 = \frac{1}{2} | \epsilon^{(+)}_\mu j^\mu_{em} | \psi_{N^*}, s_3 = -\frac{1}{2} \right\rangle e^{i|\vec{k}|x^3}, \tag{4.3.50}$$

$$S_{1/2}(Q^2) = \sqrt{\frac{2\pi\alpha}{K}} \int d^3x \left\langle \psi_N, s_3 = \frac{1}{2} | \frac{|\vec{k}|}{Q} \epsilon^{(0)}_\mu j^\mu_{em} | \psi_{N^*}, s_3 = \frac{1}{2} \right\rangle e^{i|\vec{k}|x^3}. \tag{4.3.51}$$

Here, α is the fine structure constant, Q the virtuality of the photon, and the 3-momentum $\vec{k}$ of the photon is assumed to be directed along the x^3 axis in the N^* rest frame. Also, $A_{3/2}$ is defined only when considering the transition from a resonant

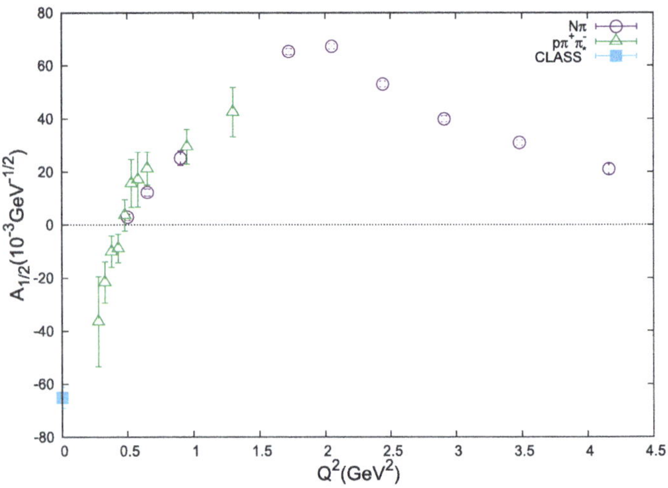

Fig. 4.2 The transverse helicity amplitude $A_{1/2}$ in units of 10^{-3} GeV$^{-1/2}$ as function of the four-momentum transfer Q^2. The sources of experimental data are shown in the panel [21–25]

state with spin 3/2 to a nucleon. Due to the energy conservation law, we have the following equation:

$$k^2 = Q^2 + \frac{(Q^2 + m_i^2 - m_f^2)^2}{4m_f^2}. \tag{4.3.52}$$

In the case of real photons, i.e., $Q^2 = 0$, we find that $|\vec{k}|$ becomes $K = (m_f^2 - m_i^2)/(2m_f)$. The polarization vectors for the axis are defined by

$$\epsilon_\mu^{(0)} = \frac{1}{Q}(|\vec{k}|, 0, 0, -k_0) \quad \text{(longitudinal mode)} \tag{4.3.53}$$

$$\epsilon_\mu^{(\pm)} = \frac{1}{\sqrt{2}}(0, 1, \pm i, 0) \quad \text{(transverse mode)}. \tag{4.3.54}$$

4.3.2.2 Results

In this subsection, we determine two parameters Kaluza-Klein Mass M_{KK} and κ, which we determine according to Ref. [2] as follows;

$$M_{KK} = 488 \text{ MeV} \tag{4.3.55}$$

$$\kappa = 0.0243. \tag{4.3.56}$$

■ Roper resonance

Here we substitute (4.1.33), (4.1.34) and (4.1.38) into the electromagnetic current expression,

$$j^{\mu}_{em} = j^{\mu,C=3}_{V} + \frac{1}{N_c} j^{\mu,C=0}_{V}, \tag{4.3.57}$$

then we get the following concrete expression [2];

$$j^0_{em}(x^{\mu}) = \frac{3}{4\pi}\frac{\rho^2}{(r^2+\rho^2)^{5/2}} I_3 + \frac{15}{16\pi}\frac{\rho^4}{(r^2+\rho^2)^{7/2}} \tag{4.3.58}$$

$$\begin{aligned} j^i_{em}(x^{\mu}) =& \frac{4\pi\kappa}{\rho^2}\Big(\frac{8}{r} - \frac{8r^4+20\rho^2 r^2+15\rho^4}{(r^2+\rho^2)^{5/2}}\Big)\epsilon_{ijk}x^j \mathrm{tr}(t_k W^{-1} t_3 W) \\ &+ \frac{15}{32\pi}\frac{\rho^4}{(r^2+\rho^2)^{7/2}}\Big(-\epsilon_{ija}x^j \chi^a + 2x^i \frac{d}{dt}\ln\rho\Big) \end{aligned} \tag{4.3.59}$$

where I_a is isospin operator $I_a = 8\pi^2\kappa\rho^2 \mathrm{tr}\big(i\dot{W}W^{-1}t_a\big)$ and $i, a = 1, 2, 3$. Here, in the present case, the term of the time derivative of Z is omitted because it is zero due to the evenness of the wave function.

Using the wave function (2.4.121), (2.4.122), and the concrete expression of the current (4.3.58) and (4.3.59), the helicity amplitude is calculated as follows

$$\begin{aligned} A_{1/2}(Q^2) =& -\sqrt{\frac{2\pi\alpha}{K}}\int_0^{\infty} d\rho\rho^3 \psi_{n_\rho=0}(\rho)\psi_{n_\rho=1}(\rho) \\ & \int_0^{\infty} drr^3 \Big\{\frac{16\pi^2\kappa}{3\sqrt{2}\rho^2}\Big(\frac{8}{r} - \frac{8r^4+20r^2\rho^2+15\rho^4}{(r^2+\rho^2)^{5/2}}\Big) + \frac{15}{32\sqrt{2}\pi^2\kappa}\frac{\rho^2}{(r^2+\rho^2)^{7/2}}\Big\} j_1(kr), \end{aligned} \tag{4.3.60}$$

$$\begin{aligned} S_{1/2}(Q^2) =& -\sqrt{\frac{2\pi\alpha}{K}}\int_0^{\infty} d\rho\rho^3 \psi_{n_\rho=0}(\rho)\psi_{n_\rho=1}(\rho) \\ & \int_0^{\infty} drr^2 \Big(\frac{3}{2}\frac{\rho^2}{(r^2+\rho^2)^{5/2}} + \frac{15}{4}\frac{\rho^4}{(r^2+\rho^2)^{7/2}}\Big) j_0(kr), \end{aligned} \tag{4.3.61}$$

where $j_n(x)$ is a spherical Bessel function, the normalized wave function ψ^l_{ρ} is defined by

$$\psi_{n_\rho}(\rho) = \frac{R_{n_\rho}(\rho)}{\sqrt{\int_0^{\infty} d\rho\rho^3 R^2_{n_\rho}(\rho)}}, \tag{4.3.62}$$

with the confluent hypergeometric functions R_{n_ρ} (2.4.94) and for the calculation of $S_{1/2}$, we used j^0_{em} from the current coservation law $Q_{\mu} j^{\mu}_{em} = 0$.

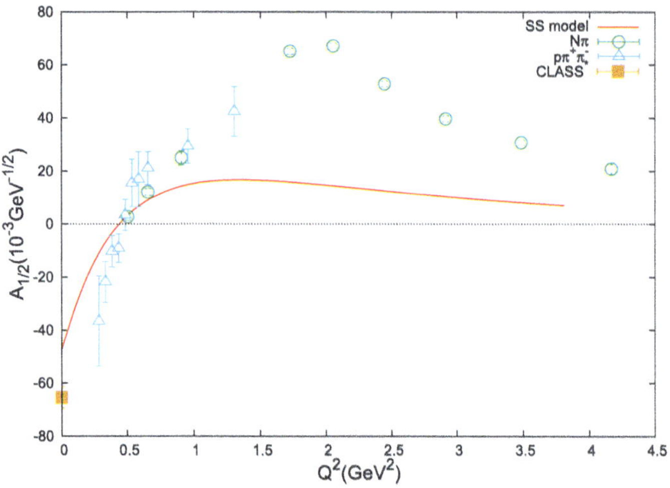

Fig. 4.3 The transverse helicity amplitude $A_{1/2}$ in units of 10^{-3} GeV$^{-1/2}$ as function of the four-momentum transfer Q^2 (Reprinted from [29], by the Authors licensed under Creative Commons Attribution 4.0 International license). The red line is result of the present study, and the sources of experimental data (CLAS) are shown in the panel [21–25]

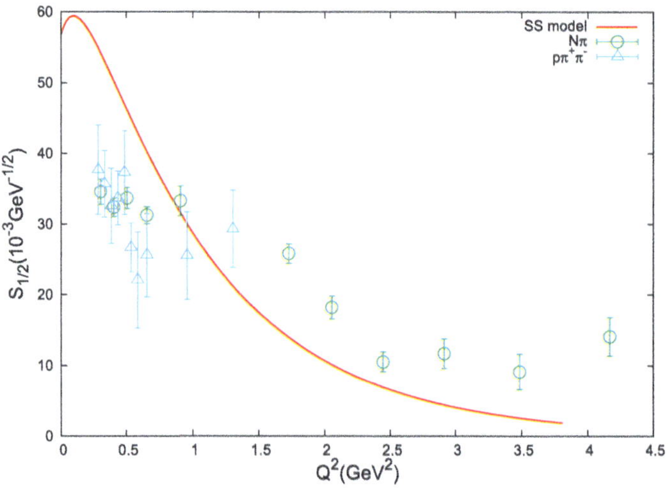

Fig. 4.4 The longitudinal helicity amplitude $S_{1/2}$ in units of 10^{-3} GeV$^{-1/2}$ (Reprinted from [29], by the Authors licensed under Creative Commons Attribution 4.0 International license). The same conventions are used as in Fig. 4.3

By performing numerical integration, the helicity amplitudes of the electromagnetic transition for the Roper resonance obtained from our calculation are shown in Figs. 4.3 and 4.4. It can be seen that this result achieves global agreement with experimental data. These results have some remarkable properties, as follows.

(i) We find that our $A_{1/2}$ near the real photon point $Q^2 = 0$ has a finite negative value. The non-relativistic quark model fails to explain this property. This is due to the orthogonality between the radial wave function of the Roper resonance and the ground state nucleon obtained from the quark model, which leads to the transition process mentioned to be a forbidden process. Several theoretical studies have attempted to solve this problem and have discussed the importance of relativistic corrections and the effect of meson clouds. The present approach shows that, in addition to the above, it is important to take into account the collective motion resulting from the soliton picture of the baryon. An important result of the collective motion comes from a dependence the current on the instanton size ρ. When we expand the current by powers of ρ, its leading term starts from ρ^2. This is an essential property of the soliton picture of baryons, indicating that the collective motion of mesons plays an important role. This picture is quite different from the single-particle picture of the quark model. Our results indicate that the collective motion mechanism of baryons is an important contribution to our understanding of the phenomena associated with Roper resonances (excitation energies and decays).
(ii) The experimental data for $A_{1/2}$ flips its sign around $Q^2 \sim 0.5\,\text{GeV}^2$. Our results roughly capture this behavior around $Q^2 \sim 0.7\,\text{GeV}^2$.
(iii) Our model prediction underestimates the experimental data of $A_{1/2}$ at $Q^2 \gtrsim 1\,\text{GeV}^2$. This is because our results are calculated up to order $1/\lambda$. Moreover, the model by meson fields should be applied to the low energy region $Q^2 \leq 1\,\text{GeV}^2$. Also, for energy regions larger than M_{KK}, the contribution from redundant modes not present in QCD, which should have been decoupled in Sect. 2.2, is larger, making the prediction less reliable.
(iv) Our prediction closely approximates the experimental data for $S_{1/2}$ although there is some overestimate. In the calculation of $S_{1/2}$, we used only the time component of the current (4.3.58) because we used the current conservation law. This charge density satisfies

$$\int d^3x j^0_{em} = I_3 + \frac{1}{2}, \tag{4.3.63}$$

for the electric charge. We consider that the fact that our prediction of $S_{1/2}$ reproduces the experimental data well is ensured by the fact that it satisfies (4.3.63).
(v) There are only two parameters in this model, κ and M_{KK}. We determined these parameters from the mass of the nucleon and the delta. With this set of parameters, the static nature of the baryons has been studied, and it has been found to predict the experimental data well (Sect. 4.2). We emphasize that the

tuning of only two parameters predicts well not only the static but even the dynamical properties of baryons.

■ $N^*(1535)$

In the calculation of the helicity amplitude of the negative parity resonance $N^*(1535)$, when using (4.3.58) and (4.3.59) as the electromagnetic current, it becomes zero due to the wave function evenness of Z. Therefore, it is necessary to newly consider the time derivative term of Z, which was ignored in the analysis of the Roper resonance. The electromagnetic current obtained is as follows,

$$j^0_{\mathrm{em}} = \frac{4\pi\kappa}{\rho^2}\left[\left(\frac{8}{r} - \frac{8r^4 + 20\rho^2 r^2 + 15\rho^4}{(r^2+\rho^2)^{5/2}}\right)\right]\dot{Z}x^b \mathrm{tr}(Wt_bW^{-1}t_a), \quad (4.3.64)$$

where the spatial component of the current contributing to $A_{1/2}$ of the $N^*(1535)$ becomes zero when only the leading of the $1/\lambda$ expansion is considered. Therefore, by considering the effect of the warp factor and the spatial component of the U(1) gauge fields, the current spatial component is

$$\begin{aligned} j^i_{\mathrm{em}}(t,x) = \frac{8\pi\kappa}{\rho^4}\Bigg[& \left(\frac{8r^2(4r^2+3\rho^2)}{\sqrt{r^2+\rho^2}} + \frac{\rho^6}{(r^2+\rho^2)^{3/2}} - 8r(4r^2+\rho^2)\right) Z\mathrm{tr}(Wt_cW^{-1}t_a) \\ & + \frac{4}{3}\left(\frac{4(6r^2+\rho^2)}{r} - \frac{24r^4+40\rho^2r^2+15\rho^4}{(r^2+\rho^2)^{3/2}}\right) x^i x^c Z\mathrm{tr}(Wt_cW^{-1}t_a)\Bigg], \end{aligned} \quad (4.3.65)$$

when the calculation is performed up to subleading of the $1/\lambda$ expansion. Here, we still see that the spatial component of U(1) gauge fields does not contribute to $A_{1/2}$. From the above, the helicity amplitudes can be calculated as follows.

$$\begin{aligned} A_{1/2} = & \sqrt{\frac{2\pi\alpha}{K}}\frac{32\pi^2\kappa}{3}\int_0^\infty d\rho\rho^3\psi_{n_\rho=0}(\rho)^2 \int_{-\infty}^{\infty} dZZ^2\psi_{n_Z=0}(Z)\psi_{n_Z=1}(Z) \\ & \int_0^\infty drr^2\Bigg[\left(\frac{8r^2(4r^2+3\rho^2)}{\sqrt{r^2+\rho^2}} + \frac{\rho^6}{(r^2+\rho^2)^{3/2}} - 8r(4r^2+\rho^2)\right) j_0(kr) \\ & -\frac{4}{3}\left(4r(6r^2+\rho^2) - \frac{r^2(24r^4+40\rho^2r^2+15\rho^4)}{(r^2+\rho^2)^{3/2}}\right)\frac{j_1(kr)}{kr}\Bigg], \end{aligned} \quad (4.3.66)$$

$$\begin{aligned} S_{1/2} = & -\sqrt{\frac{2\pi\alpha}{K}}\int_0^\infty d\rho\rho^3\psi_{n_\rho=0}(\rho)^2 \\ & \int_0^\infty drr^2\left[\frac{8\pi\kappa r\rho^2}{3\sqrt{6}}\left(\frac{8}{r} - \frac{8r^4+20\rho^2r^2+15\rho^4}{(r^2+\rho^2)^{3/2}}\right)\langle Z^2\rangle\right] j_1(kr), \end{aligned} \quad (4.3.67)$$

where, note the normalization, $\langle Z^2\rangle$ is the expectation value of the wave function of Z, and using

$$\left\langle n_Z = 1 \left| \left(-\frac{i}{M_0}\frac{\partial}{\partial Z} \right) \right| n_Z = 0 \right\rangle = \frac{\int_{-\infty}^{\infty} dZ Z e^{-\frac{M_0}{\sqrt{6}}Z^2} \left(-\frac{i}{M_0}\frac{\partial}{\partial Z} \right) e^{-\frac{M_0}{\sqrt{6}}Z^2}}{\sqrt{\int_{-\infty}^{\infty} dZ Z^2 e^{-\frac{2M_0}{\sqrt{6}}Z^2} \int_{-\infty}^{\infty} dZ e^{-\frac{2M_0}{\sqrt{6}}Z^2}}}$$

$$= \frac{\int_{-\infty}^{\infty} dZ Z e^{-\frac{M_0}{\sqrt{6}}Z^2} \left(-\frac{2i}{\sqrt{6}} Z \right) e^{-\frac{M_0}{\sqrt{6}}Z^2}}{\int_{-\infty}^{\infty} dZ Z^2 e^{-\frac{2M_0}{\sqrt{6}}Z^2} \int_{-\infty}^{\infty} dZ e^{-\frac{2M_0}{\sqrt{6}}Z^2}} = -\frac{2i}{\sqrt{6}} \frac{\int_{-\infty}^{\infty} dZ e^{-\frac{2M_0}{\sqrt{6}}Z^2}}{\int_{-\infty}^{\infty} dZ Z^2 e^{-\frac{2M_0}{\sqrt{6}}Z^2}} \langle n_Z = 0 | Z^2 | n_Z = 0 \rangle$$

$$= -\frac{2i}{\sqrt{6}} \frac{32\pi^2\kappa}{\sqrt{6}} \langle Z^2 \rangle, \tag{4.3.68}$$

with (2.4.111).

We have performed numerical computations of the integral (4.3.66) and (4.3.67) and show the helicity amplitudes of the electromagnetic transition amplitudes for the negative parity resonance $N^*(1535)$, Figs. 4.5 and 4.6. Some comments on these results are given below.

(i) For $A_{1/2}$, there is no contribution from leading of $1/\lambda$ expansion. It is also found that the calculation with the contribution of subleading underestimates the experimental data. We consider that this is because this calculation is based on the $1/\lambda$ expansion. It is possible to evaluate the current without the $1/\lambda$ expansion. It is important to consider the contribution of the warp factor to the classical solution, because the excitation of the negative parity state corresponds to the oscillation in the Z direction. By the same argument as in Sect. 4.1.2, it is possible to obtain an asymptotic solution for the instanton configuration. Substituting this solution into the 5d Noether current yields a new contribution from the $\tilde{j}^{\mu}_{L/R}(x)$ term in Eq. (4.1.109). This term is essentially the same as

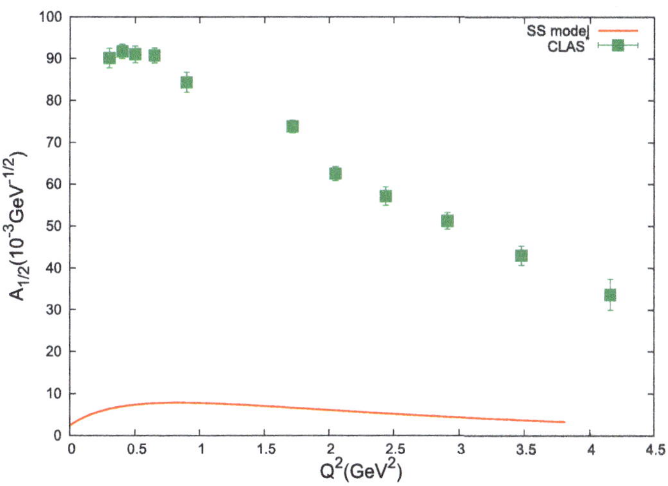

Fig. 4.5 The transverse helicity amplitude $A_{1/2}$ in units of 10^{-3} $\text{GeV}^{-1/2}$ as function of the four-momentum transfer Q^2. The red line is result of the present study, and the sources of experimental data are shown in the panel [22]

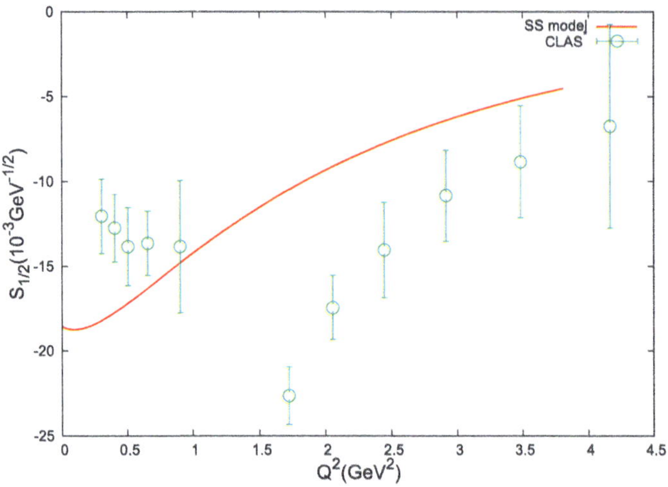

Fig. 4.6 The transverse helicity amplitude $S_{1/2}$ in units of 10^{-3} GeV$^{-1/2}$ as function of the four-momentum transfer Q^2. The red line is result of the present study, and the sources of experimental data are shown in the panel [22]

the contribution coming from the GKP-Witten current. The difference in the gauge configuration of the instanton solution appears in the difference in the magnitude of the coupling constant of the meson-baryon composite system in four dimensions. As explained in Sect. 1.1, the Negative parity resonance is known to be strongly coupled to ηN. Therefore, it is necessary to consider a gauge-field configuration that reproduces the coupling properties of the negative parity resonance to the meson. From this point of view, it is also necessary to consider the effect of the warp factor to obtain a classical solution. This is a future challenge, which, if accomplished, is expected to further deepen understanding of holographic QCD dynamics.

(ii) It can be seen that the shape of $A_{1/2}$ is very similar to the experimental data, even though the absolute values are quite small. In particular, the peak value is $Q^2 = 0.6$ GeV^2, which is very close to the experimental behavior.

(iii) For $S_{1/2}$, there is a leading contribution of $1/\lambda$ expansion, indicating that it reproduces the experimental data very well.

(iv) The 5d Noether current has a non-uniqueness in the determination of $\zeta(x, z)$, which has been determined to reproduce the chiral current of the Skyrme model. However, in the Skyrme model it is not easy to generate negative parity states within the collective quantization method; one way is to introduce meson fluctuations around the soliton solutions [30]. To take into account such dynamics, it may be necessary to reconsider the criteria for determining $\zeta(x, z)$, which is however beyond the present study.

■ $\Delta(1232)$

Calculate the helicity amplitude of the Delta resonance $\Delta(1232)$. For $A_{3/2}$ and $A_{1/2}$, we can use the current of (4.3.59). As for $S_{1/2}$, we can see that if we define it using j^0, the value will vanish in the leading of $1/\lambda$ expansion, so we consider the effect of the warp factor, the time component of current is

$$j^0_{\rm em}(x) = \frac{1}{6\pi\rho^4}\Big[\left(\frac{8}{r} - \frac{8r^4+20\rho^2r^2+15\rho^4}{(r^2+\rho^2)^{5/2}}\right)Z^2 - \left(8r - \frac{8r^4+12\rho^2r^2+3\rho^4}{(r^2+\rho^2)^{3/2}}\right)\Big](x^3)^2 J_3 \mathrm{tr}(Wt_3W^{-1}t_3). \quad (4.3.69)$$

From the above, the helicity amplitudes are written as

$$\begin{aligned} A_{1/2} &= -\sqrt{\frac{2\pi\alpha}{K}}\int_0^\infty drr^2\int_0^\infty d\rho\rho^3\psi_{n_\rho=0}(\rho)^2 j_1(kr) \\ &= \frac{8\pi^2\kappa}{3\rho^2}\left(\frac{8}{r} - \frac{8r^4+20\rho^2r^2+15\rho^4}{(r^2+\rho^2)^{5/2}}\right) \end{aligned} \quad (4.3.70)$$

$$\begin{aligned} A_{3/2} &= -\sqrt{\frac{2\pi\alpha}{K}}\int_0^\infty drr^2\int_0^\infty d\rho\rho^3\psi_{n_\rho=0}(\rho)^2 j_1(kr) \\ &= \frac{8\sqrt{2}\pi^2\kappa}{3\rho^2}\left(\frac{8}{r} - \frac{8r^4+20\rho^2r^2+15\rho^4}{(r^2+\rho^2)^{5/2}}\right) \end{aligned} \quad (4.3.71)$$

$$\begin{aligned} S_{1/2} &= \sqrt{\frac{2\pi\alpha}{K}}\int_0^\infty drr^2\int_0^\infty d\rho\rho^3\psi_{n_\rho=0}(\rho)^2\frac{2\sin kr - (kr)^2\sin kr - 2kr\cos kr}{(kr)^3} \\ &\times\frac{\sqrt{2}r^2}{18\rho^4}\Big[\langle Z^2\rangle\left(\frac{8}{r} - \frac{8r^4+20\rho^2r^2+15\rho^4}{(r^2+\rho^2)^{5/2}}\right) - \left(8r - \frac{8r^4+12\rho^2r^2+3\rho^4}{(r^2+\rho^2)^{3/2}}\right)\Big]. \end{aligned} \quad (4.3.72)$$

We performed numerical computations of the integral (4.3.70), (4.3.71) and (4.3.72) and show the helicity amplitudes of the electromagnetic transition amplitudes for the delta resonance $\Delta(1232)$, Figs. 4.7, 4.8 and 4.9. Some comments on these results are given below.

(i) For $A_{1/2}$, there is a leading contribution of $1/\lambda$ expansion. However, the obtained values are considerably smaller than the experimental data, although they reproduce the approximate behavior. We consider that this is because the present calculation follows the prescription of $1/\lambda$ expansion as well as Roper resonance and Negative parity resonance.

(ii) For $S_{1/2}$, there is no leading contribution from the $1/\lambda$ expansion. Therefore, we performed a calculation of the subleading of the $1/\lambda$ expansion including the effect of the warp factor. However, the absolute values obtained are much smaller than the experimental data.

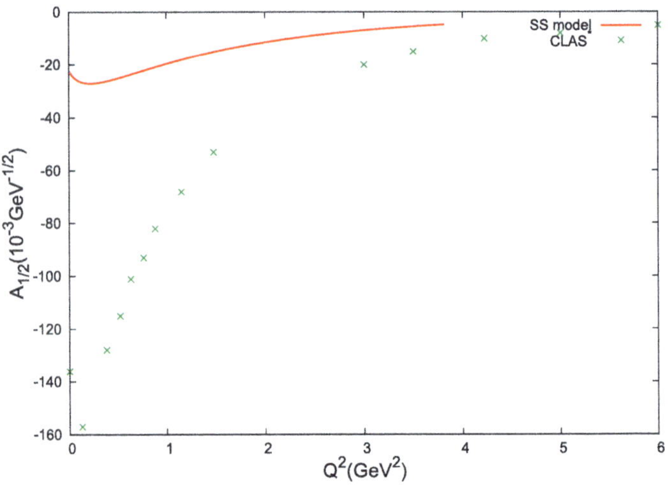

Fig. 4.7 The transverse helicity amplitude $A_{1/2}$ in units of 10^{-3} GeV$^{-1/2}$ as function of the four-momentum transfer Q^2. The red line is result of the present study, and the sources of experimental data are shown in the panel [22, 31]

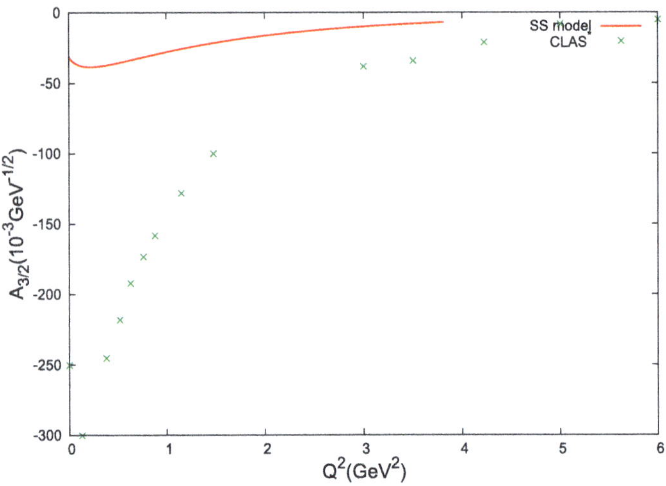

Fig. 4.8 The transverse helicity amplitude $A_{3/2}$ in units of 10^{-3} GeV$^{-1/2}$ as function of the four-momentum transfer Q^2. The red line is result of the present study, and the sources of experimental data are shown in the panel [22, 31]

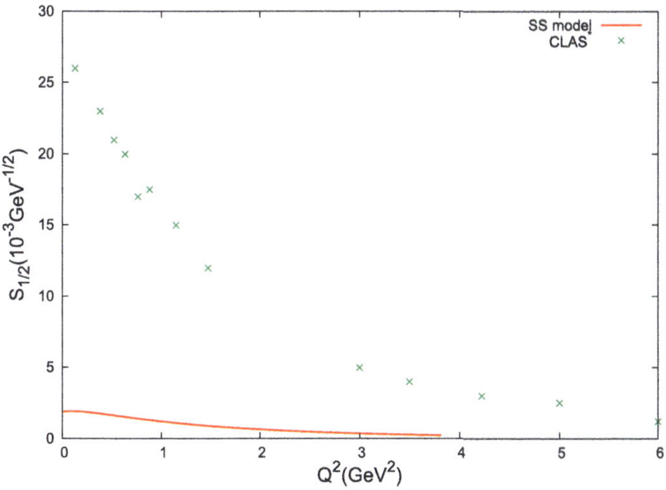

Fig. 4.9 The transverse helicity amplitude $S_{1/2}$ in units of 10^{-3} GeV$^{-1/2}$ as function of the four-momentum transfer Q^2. The red line is result of the present study, and the sources of experimental data are shown in the panel [22, 31]

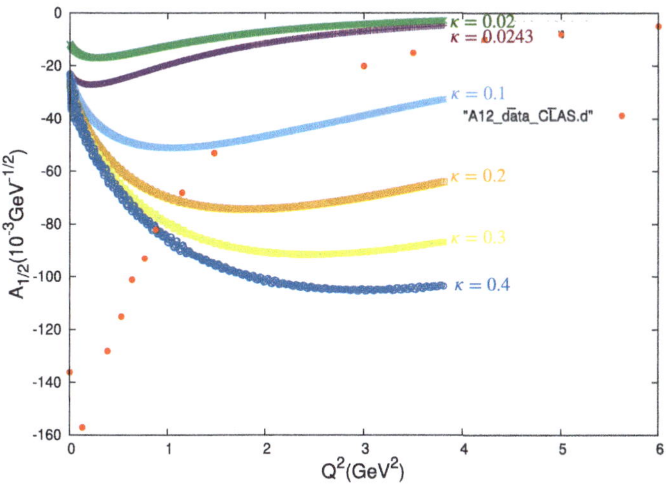

Fig. 4.10 the κ dependence of the numerical results of the transverse helicity amplitude $A_{1/2}$ in units of 10^{-3} GeV$^{-1/2}$ as function of the four-momentum transfer Q^2

(iii) The results of our analysis depend significantly on the parameter κ of this model. We show the κ dependence of the numerical results in Fig. 4.10. We can see that the absolute value of our prediction of the electromagnetic transition amplitude becomes larger as κ becomes larger. However, as κ increases, we overestimate the experimental data in the high-momentum transition region.

4.3.2.3 Discussion

In this subsection, we have performed an analysis of the electromagnetic form factor of a typical nucleon resonance using the 5d Noether current. As a result, it is shown that $S_{1/2}$ of the Roper resonance and the negative parity resonance approximately reproduce the experimental data. In particular, for the Roper resonance $A_{1/2}$, we find that at the real photon point $Q^2 = 0$, it reproduces an interesting feature that takes a finite negative value, which the quark model fails to explain. This is essentially owing to the representation as a collective motion of the nucleon resonance. The representation of baryon resonances is completely different from the quark model, which considers quark three-body dynamics, suggesting the importance of the collective motion in the excitation created by the QCD vacuum. In Fig. 4.11, we compare our prediction with the non-relativistic quark model. Our prediction shows a significant improvement compared to that of the quark model concerning the reproduction of the experimental data.

On the other hand, our results underestimate the experimental data overall for the $A_{1/2}$ of the negative parity resonance and Delta resonance $\Delta(1232)$. However, the

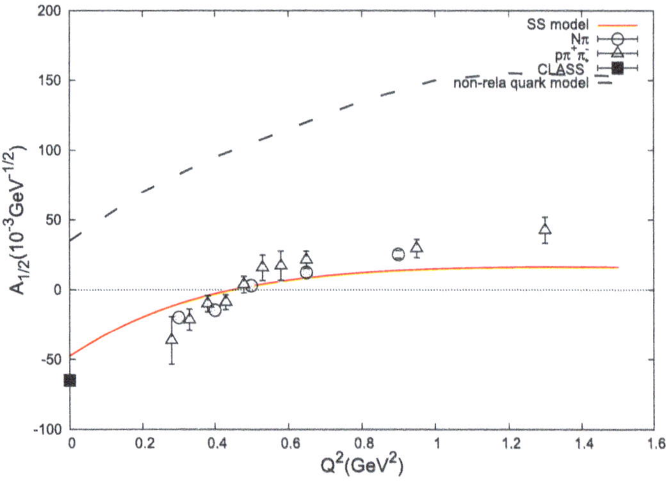

Fig. 4.11 The transverse helicity amplitude $A_{1/2}$ in units of 10^{-3} GeV$^{-1/2}$ as function of the four-momentum transfer Q^2. The red line is result of the present study, and the sources of experimental data are shown in the panel [21–25]

results we have given in this subsection are calculations based on the $1/\lambda$ expansion. Therefore, we expect an improvement in our results by performing the calculations without the $1/\lambda$ expansion. This method is discussed in Ref. [1]. We expect that our results will be further improved by using their asymptotic solutions (4.1.68) and (4.1.69).

The 5d Noether current can be used to obtain the axial coupling obtained in Sect. 4.2.4. By using the current (4.2.59) defined in Sect. 4.1.3.1, the axial coupling of nucleon is written by,

$$\begin{aligned}\frac{1}{2}g_A^{NN} &= \int d^3x j_A^{i,C=i}(x^\mu) \\ &= \int d^3x \frac{32\kappa}{3\rho^2}\left(8r - \frac{8r^4 + 12\rho^2 r^2 + 3\rho^4}{(r^2+\rho^2)^{3/2}}\right)\mathrm{tr}(t_i W^{-1} t_a W) \\ &= -\frac{32\pi\kappa\rho^2}{3}\mathrm{tr}(t_i W^{-1} t_a W). \end{aligned} \tag{4.3.73}$$

By performing the Fourier transform, we obtain the axial coupling;

$$\begin{aligned} g_A^{NN}(\vec{k} \to 0) &= \frac{16\pi\kappa}{3}\langle\psi_N|\rho^2|\psi_N\rangle \\ &= \frac{\sqrt{6}}{3\pi}\left(1 + 2\sqrt{1+\frac{N_c^2}{5}}\right) = 1.13, \end{aligned} \tag{4.3.74}$$

where $\vec{k}$ is 3-momentum of pion. It is interesting to note that the axial coupling constant obtained from this current is a value that does not depend on the parameters of our model. In the same way as in Sect. 4.3.1 there is also a relation between the axial coupling g_A^{NN} (4.3.74) and the axial coupling of the $N^*(1440) \to N + \pi$ transition that does not depend on the parameters of our model as follows;

$$g_A^{NN}/g_A^{NN^*} = \left(1 + 2\sqrt{1+\frac{N_c^2}{5}}\right)^{1/2} = 2.08. \tag{4.3.75}$$

From this relation, we obtain $g_A^{NN^*} = 0.543$. According to Refs. [17, 32], the decay width of the $N^*(1440) \to N + \pi$ transition is

$$\begin{aligned} &\Gamma_{N^*(1440)\to N+\pi} \\ &= \frac{k}{4\pi}\frac{M_N + E_N}{M_{N^*}}\left(\frac{M_N + M_{N^*}}{2f_\pi}\frac{g_A^{NN^*}q}{E_N + M_N}\right)^2. \end{aligned} \tag{4.3.76}$$

Using $M_N = 940$ MeV, $M_{N^*} = 1370$ MeV, $k = 342$ MeV, $m_\pi = 140$ MeV (pion mass), $f_\pi = 64.5$ MeV (pion decay constant), we find

$$\Gamma_{N^*(1440)\to N+\pi} = 117\ \mathrm{MeV}, \tag{4.3.77}$$

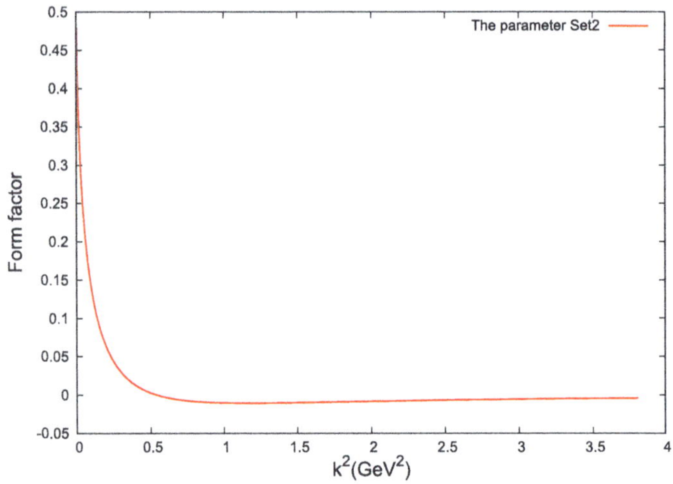

Fig. 4.12 The emitted pion momentum $\boldsymbol{k}$ dependence of the form factor

where, in this computation, the value of $g_A^{NN^*}$ at $\vec{k} = 0$ is used. These results provide quite good values for the experimental data. However, the shape of the form factor shows a sharp decrease with increasing $\vec{k}$. Therefore, when the effect of the form factor is taken into account, the value is considerably smaller than the experimental value (Fig. 4.12). On this point, again, the 5d Noether current does not explain the behavior well.

As described above, there are several problems when analyzing the dynamical properties of nucleon resonance using the 5d Noether current. Nevertheless, it is noteworthy that we can reproduce the experimental data of $A_{1/2}$ of the Roper resonance. It is difficult to explain the experimental data of $A_{1/2}$ (Roper) in the quark model. Therefore, our results indicate the importance of the collective motion mechanism of the mesons that construct the nucleon.

4.3.3 Summary and Discusstion

In this section, we investigate the dynamical properties of nucleons by using the current defined by the Sakai-Sugimoto model. There is a problem with the definition of current in the Sakai-Sugimoto model. The solution to this problem is discussed from two viewpoints [1, 2], leading to the GKP-Witten current and the 5d Noether current, respectively. First, using the GKP-Witten current, we calculated the axial coupling of the nucleon resonance and the decay width of the one pion emission derived from it, showing that it reproduces the experimental data well. This suggests that the effect derived from the collective motion of baryons/solitons in this model is important. On the other hand, turning to the electromagnetic transition amplitudes, we find that the GKP-Witten current cannot reproduce the experimental data for $A_{1/2}$ of the Roper resonance. Therefore, we next calculated the electromagnetic transition amplitude of the nucleon resonance using the 5d Noether current and compared it with the experimental data. As a result, we succeeded in capturing the features of the experimental data of $A_{1/2}$ of the Roper resonance well. However, it fails to reproduce the experimental data of the electromagnetic transition amplitudes of the Negative parity resonance and the Delta resonance. Moreover, while it reproduces the value of axial coupling well, the momentum dependence of its form factor is inconsistent with the experimental data. This may indicate a limitation of the analysis by $1/\lambda$ expansion. In Ref. [1], an analysis method beyond the $1/\lambda$ expansion is presented, which is expected to improve the results by performing the same analysis using an asymptotic solution for the instanton solution.

Despite the above problems, our analysis suggests that the collective motion mechanism of the meson field plays an important role in a comprehensive understanding of the nucleon resonance created by the QCD vacuum, for which further developments are expected. We need to clearly understand the role of each of the several mechanisms (for example, collective motion, single particle picture, and so on) for the description of nucleon resonances in future studies.

4.4 Appendix: The Wave Function of the SU(2) Rotation $W(t)$

4.4.1 $W(t)$

The SU(2) rotation $W(t)$ is written by

$$W = a_4(t) + i a_a(t)\tau^a, \tag{4.4.1}$$

where, the group manifold SU(2) is parameterized by

$$a_4 = \cos\theta_0, \tag{4.4.2}$$
$$a_3 = \sin\theta_0 \cos\theta_1, \tag{4.4.3}$$
$$a_2 = \sin\theta_0 \sin\theta_1 \cos\theta_2, \tag{4.4.4}$$
$$a_1 = \sin\theta_0 \sin\theta_1 \sin\theta_2, \tag{4.4.5}$$

$$d\Omega_3 = \sin^2\theta_0 \sin\theta_1 d\theta_0 d\theta_1 d\theta_2. \tag{4.4.6}$$

The spin and isospin are expressed as follows;

$$J_i = M_0\rho^2 \mathrm{tr}(-iW^{-1}\dot{W}t_i), \tag{4.4.7}$$
$$I_a = M_0\rho^2 \mathrm{tr}(i\dot{W}W^{-1}t_a) = -WJ_it_iW^{-1}. \tag{4.4.8}$$

We use the canonical momentum

$$\Pi_I = 2M_0\dot{y}_I = -i\frac{\partial}{\partial y_I}, \tag{4.4.9}$$

then the spin and isospin operator are rewritten by

$$I_a = \frac{i}{2}\left(a_4\frac{\partial}{\partial a_a} - a_a\frac{\partial}{\partial a_4} - \epsilon_{abc}a_b\frac{\partial}{\partial a_c}\right) \tag{4.4.10}$$

$$J_a = \frac{i}{2}\left(-a_4\frac{\partial}{\partial a_a} + a_a\frac{\partial}{\partial a_4} - \epsilon_{abc}a_b\frac{\partial}{\partial a_c}\right), \tag{4.4.11}$$

where $y_I = \rho a_I$.

4.4.2 *Wave Function for* $W(t)$

Here, the wave function of $W(t)$ is written, for example

$$|l = 1, I_3 = J_3 = l/2\rangle = \pi^{-1}(a_1 + ia_2), \tag{4.4.12}$$

$$|l = 3, I_3 = J_3 = l/2\rangle = \frac{\sqrt{2}}{\pi}(a_1 + ia_2)^3, \tag{4.4.13}$$

where the normalization constants were determined as follows,

$$\begin{aligned}\int d\Omega_3\pi^{-2}(a_1 - ia_2)(a_1 + ia_2) &= \int d\Omega_3\pi^{-2}(a_1^2 + a_2^2)\\ &= \int d\Omega_3\pi^{-2}(\sin^2\theta_0\sin^2\theta_1) = 1\end{aligned} \tag{4.4.14}$$

$$\begin{aligned}\int d\Omega_3\frac{2}{\pi^2}(a_1 - ia_2)^3(a_1 + a_2)^3 &= \int d\Omega_3\frac{2}{\pi^2}(a_1^2 + a_2^2)^3\\ &= \int d\Omega_3\frac{2}{\pi^2}(\sin^2\theta_0\sin^2\theta_1)^3 = 1\end{aligned} \tag{4.4.15}$$

With ladder operator

$$I_- = I_1 - iI_2, \quad I_- = I_1 + iI_2 \tag{4.4.16}$$

$$J_- = J_1 - iJ_2, \quad J_- = J_1 + iJ_2, \tag{4.4.17}$$

we obtain several baryon states

$$\begin{aligned}
&\left|I=1, I_3=\frac{1}{2}, J_3=-\frac{1}{2}\right\rangle = J_-(a_1+ia_2)\\
&=\Big[\frac{i}{2}\Big(-a_4\frac{\partial}{\partial a_1}+a_1\frac{\partial}{\partial a_4}-a_2\frac{\partial}{\partial a_3}+a_3\frac{\partial}{\partial a_2}\Big)\\
&\quad+\frac{1}{2}\Big(-a_4\frac{\partial}{\partial a_2}+a_2\frac{\partial}{\partial a_4}+a_1\frac{\partial}{\partial a_3}-a_3\frac{\partial}{\partial a_1}\Big)\Big](a_1+ia_2)\\
&=-i\pi^{-1}(a_4-ia_3)
\end{aligned}$$

$$\left|I=1, I_3=-\frac{1}{2}, J_3=\frac{1}{2}\right\rangle = I_-\left|I=1, I_3=\frac{1}{2}, J_3=\frac{1}{2}\right\rangle = i\pi^{-1}(a_4+ia_3) \tag{4.4.18}$$

$$\left|I=1, I_3=-\frac{1}{2}, J_3=-\frac{1}{2}\right\rangle = J_-\left|I=1, I_3=\frac{1}{2}, J_3=\frac{1}{2}\right\rangle = -\pi^{-1}(a_1-ia_2) \tag{4.4.19}$$

$$\left|I=3, I_3=\frac{3}{2}, J_3=\frac{1}{2}\right\rangle = J_-\left|I=3, I_3=\frac{3}{2}, J_3=\frac{3}{2}\right\rangle = -\frac{\sqrt{6}}{\pi}i(a_1+ia_2)^2(a_4-ia_3) \tag{4.4.20}$$

$$\left|I=3, I_3=\frac{1}{2}, J_3=\frac{3}{2}\right\rangle = I_-\left|I=3, I_3=\frac{1}{2}, J_3=\frac{3}{2}\right\rangle = \frac{\sqrt{6}}{\pi}i(a_1+ia_2)^2(a_4+ia_3) \tag{4.4.21}$$

$$\begin{aligned}
&\left|I=3, I_3=\frac{1}{2}, J_3=\frac{1}{2}\right\rangle = J_-I_-\left|I=3, I_3=\frac{3}{2}, J_3=\frac{3}{2}\right\rangle\\
&=\frac{\sqrt{2}}{\pi}(a_1+ia_2)(2a_3^2+2a_4^2-a_1^2-a_2^2)
\end{aligned} \tag{4.4.22}$$

4.4.3 Some Expectation Values

We show the calculations of some expectation values as follows.

$$\mathrm{tr}(W\tau^3W^{-1}\tau^3) = 4a_4^2+4a_3^2-2 \tag{4.4.23}$$

■ $\langle N|\mathrm{tr}(W\tau^3W^{-1}\tau^3)|N\rangle$

$$\begin{aligned}
&\langle I=1, I_3=J_3=I/2|\mathrm{tr}(W\tau^3W^{-1}\tau^3)|I=1, I_3=J_3=I/2\rangle\\
&=\int d\Omega_3\pi^{-2}(a_1-ia_2)(4a_4^2+4a_3^2-2)(a_1+ia_2) = \int d\Omega_3\pi^{-2}\sin^2\theta_0\sin^2\theta_1(2-4\sin^2\theta_0\sin^2\theta_1)\\
&=-\frac{2}{3}
\end{aligned} \tag{4.4.24}$$

■ $\langle\Delta|\mathrm{tr}(W\tau^3W^{-1}\tau^3)|N\rangle$

$$\begin{aligned}
&\langle l=1, I_3=J_3=l/2|\mathrm{tr}(W\tau^3W^{-1}\tau^3)|l=3, I_3=J_3=l/2\rangle\\
&=\int d\Omega_3\pi^{-1}(a_1-ia_2)(4a_4^2+4a_3^2-2)\frac{\sqrt{2}}{\pi}(a_1+ia_2)(2a_3^2+2a_4^2-a_1^2-a_2^2)\\
&=\int d\Omega_3\pi^{-2}\sin^2\theta_0\sin^2\theta_1(2-4\sin^2\theta_0\sin^2\theta_1)(2-3\sin^2\theta_0\sin^2\theta_1)\\
&=\frac{2\sqrt{2}}{3}
\end{aligned} \tag{4.4.25}$$

$$\mathrm{tr}(W\frac{\tau^2}{2}W^{-1}\frac{\tau^3}{2})=\mathrm{tr}[(a_4+ia_a\tau^a)\frac{\tau^2}{2}(a_4-ia_b\tau^b)\frac{\tau^3}{2}]=a_2a_3-a_1a_4 \tag{4.4.26}$$

$$\mathrm{tr}(W\frac{\tau^1}{2}W^{-1}\frac{\tau^3}{2})=\mathrm{tr}[(a_4+ia_a\tau^a)\frac{\tau^1}{2}(a_4-ia_b\tau^b)\frac{\tau^3}{2}]=a_1a_3+a_2a_4 \tag{4.4.27}$$

■ $\langle\Delta^+,\frac{3}{2}|\left[\mathrm{tr}(W\tau^2W^{-1}\tau^3)-i\mathrm{tr}(W\tau^1W^{-1}\tau^3)\right]|N\rangle$

$$\begin{aligned}
&\langle l=3, I_3=3/2, J_3=1/2|\left[\mathrm{tr}(W\tau^2W^{-1}\tau^3)-i\mathrm{tr}(W\tau^1W^{-1}\tau^3)\right]|l=1, I_3=J_3=l/2\rangle\\
&=\frac{\sqrt{6}}{\pi^2}i\int d\Omega_3(a_1^2+a_2^2)^2(a_3^2+a_4^2)=\frac{\sqrt{6}}{\pi^2}i\int d\Omega_3\sin^4\theta_0\sin^4\theta_1(1-\sin^2\theta_0\sin^2\theta_1)\\
&=i\frac{1}{\sqrt{6}}
\end{aligned} \tag{4.4.28}$$

■ $\langle\Delta^+,\frac{1}{2}|\left[\mathrm{tr}(W\tau^2W^{-1}\tau^3)-i\mathrm{tr}(W\tau^1W^{-1}\tau^3)\right]|N\rangle$

$$\begin{aligned}
&\langle l=3, I_3=1/2, J_3=1/2|\mathrm{tr}(W\tau^3W^{-1}\tau^3)|l=1, I_3=1/2, J_3=-1/2\rangle\\
&=i\frac{\sqrt{2}}{\pi^2}\int d\Omega_3(a_1^2+a_2^2)(a2_3+a_4^2)(2a_3^2+2a_4^2-a_1^2-a_2^2)\\
&=i\frac{\sqrt{2}}{\pi^2}\int d\Omega_3\sin^2\theta_0\sin^2\theta_1(1-\sin^2\theta_0\sin^2\theta_1)(2-3\sin^2\theta_0\sin^2\theta_1)\\
&=i\frac{\sqrt{2}}{6}
\end{aligned} \tag{4.4.29}$$

References

1. Hashimoto K, Sakai T, Sugimoto S (2008) Holographic Baryons: static properties and form factors from gauge/string duality. Prog Theor Phys 120:1093–1137
2. Hata H, Murata M, Yamato S (2008) Chiral currents and static properties of nucleons in holographic QCD. Phys Rev D 78:086006
3. Son DT, Stephanov MA (2004) QCD and dimensional deconstruction. Phys Rev D 69:065020
4. Gubser SS, Klebanov IR, Polyakov AM (1998) Gauge theory correlators from noncritical string theory. Phys Lett B 428:105–114
5. Witten E (1998) Anti-de Sitter space and holography. Adv Theor Math Phys 2:253–291
6. Natsuume M (2015) AdS/CFT duality user guide, vol 903
7. Atiyah MF, Manton NS (1989) Skyrmions from instantons. Phys Lett B 222:438–442
8. Adkins GS, Nappi CR, Witten E (1983) Static properties of nucleons in the Skyrme model. Nucl Phys B 228:552
9. Burkert VD, Roberts CD (2019) Colloquium: roper resonance: toward a solution to the fifty year puzzle. Rev Mod Phys 91(1):011003
10. Suzuki N, Julia-Diaz B, Kamano H, Lee TSH, Matsuyama A, Sato T (2010) Disentangling the dynamical origin of p-11 nucleon resonances. Phys Rev Lett 104:042302
11. Douglas MR (1999) Branes within branes. NATO Sci Ser C 520:267–275
12. Witten E (1998) Baryons and branes in anti-de Sitter space. JHEP 07:006
13. Hata H, Sakai T, Sugimoto S, Yamato S (2007) Baryons from instantons in holographic QCD. Prog Theor Phys 117:1157
14. Skyrme THR (1962) A unified field theory of mesons and baryons. Nucl Phys 31:556–569
15. Zyla PA et al (2020) Review of particle physics. PTEP 2020(8):083C01
16. Hosaka A, Toki H (1996) Chiral bag model for the nucleon. Phys Rept 277:65–188
17. Iwanaka A, Fujii D, Hosaka A (2022) Decay properties of N(1535) in the holographic QCD. Phys Rev D 105(11):114057
18. Isgur N, Karl G (1979) Positive parity excited baryons in a quark model with hyperfine interactions. Phys Rev D 19:2653. [Erratum: Phys Rev D 23:817 (1981)]
19. Capstick S (1992) Photoproduction and electroproduction of nonstrange baryon resonances in the relativized quark model. Phys Rev D 46:2864–2881
20. Capstick S, Keister BD (1995) Baryon current matrix elements in a light front framework. Phys Rev D 51:3598–3612
21. Aznauryan IG et al (2008) Electroexcitation of the roper resonance for 1.7 < Q**2 < 4.5 -GeV2 in vec-ep —> en pi+. Phys Rev C 78:045209
22. Aznauryan IG et al (2009) Electroexcitation of nucleon resonances from CLAS data on single pion electroproduction. Phys. Rev. C 80:055203
23. Dugger M et al (2009) pi+ photoproduction on the proton for photon energies from 0.725 to 2.875-GeV. Phys Rev C 79:065206
24. Mokeev VI et al (2012) Experimental Study of the $P_{11}(1440)$ and $D_{13}(1520)$ resonances from CLAS data on $ep \to e'\pi^+\pi^- p'$. Phys Rev C 86:035203
25. Mokeev VI et al (2016) New results from the studies of the $N(1440)1/2^+$, $N(1520)3/2^-$, and $\Delta(1620)1/2^-$ resonances in exclusive $ep \to e'p'\pi^+\pi^-$ electroproduction with the CLAS detector. Phys Rev C 93(2):025206
26. Capstick S (1992) Photoproduction amplitudes of P(11) and P(33) baryon resonances in the quark model. Phys Rev D 46:1965–1972
27. Kubota T, Ohta K (1976) Relativistic corrections to the Baryon resonance photoexcitation amplitudes in the quark model. Phys Lett B 65:374–376
28. Arifi AJ, Suenaga D, Hosaka A (2021) Relativistic corrections to decays of heavy baryons in the quark model. Phys Rev D 103(9):094003
29. Fujii D, Iwanaka A, Hosaka A (2022) Electromagnetic transition amplitude for Roper resonance from holographic QCD. Phys Rev D 106(1):014010

30. Callan J, Curtis G, Klebanov IR (1985) Bound state approach to strangeness in the Skyrme model. Nucl Phys B 262:365–382
31. Drechsel D, Kamalov SS, Tiator L (2007) Unitary Isobar Model—MAID2007. Eur Phys J A 34:69–97
32. Fujii D, Hosaka A (2021) Decay properties of Roper resonance in the holographic QCD. Phys Rev D 104(1):014022

Chapter 5
Summary and Outlook

Abstract A summary of this book and future prospects for research in hadron physics using the Sakai-Sugimoto model will be given.

In this doctoral thesis, we attempted to study the dynamical properties of nucleon resonance using the Sakai-Sugimoto model. Therefore, as a preparation, in Sect. 2, we gave a brief description of the Sakai-Sugimoto model and mainly discussed how to treat mesons and baryons. In particular, we emphasized that the effective theory of hadrons derived from this model can be interpreted as a meson-baryon composite system. The Sakai-Sugimoto model leads to a 5-dimensional hadron effective theory as a holographic dual of QCD. The extra dimension then provides a description of pions, ρ-mesons and their infinite resonant states. By performing mode expansion and dimensional reduction, we obtain an action consisting of a Skyrme field as a nucleon and an infinite number of meson fields (Skyrme model and infinite number of meson fields and their couplings). This meson-baryon composite system is critically important for studying the nature of nucleon resonance. This is because the recent simple hadron models (e.g., the quark model) face some limitations with respect to the analysis of nucleon resonances, whereas models that describe nucleon resonances as resonant states of mesons and nucleons have had much success. The importance of such a description of the resonance state is also suggested from the experimental point of view. This is because the quark model describes nucleon resonances as stable particles, whereas the actually observed nucleon resonances are recognized as poles of the scattering amplitudes of mesons and baryons. We should try to understand the nature of nucleon resonance with respect to these facts.

The Sakai-Sugimoto model describes nucleon resonance as a soliton in 5-dimensional space-time consisting of meson fields. As mentioned above, this model leads to a meson-baryon composite system by dimensional reduction to 4 dimensions, so we believe that it is an interesting attempt to study the properties of nucleon resonance using this model. In particular, the Roper resonance, the first excited level of the nucleon, is considered to correspond to the monopole excitation, and the Sakai-Sugimoto model, which describes the Roper resonance as a soliton size vibration, respects this picture.

Furthermore, hadron resonances involving heavy quarks have been studied intensively in recent years. In particular, it has been shown that Roper-like excitations

D. Fujii, *Dynamical Properties of Baryon Resonances in the Holographic QCD*,
Springer Theses, https://doi.org/10.1007/978-981-97-0755-3_5

exist even in heavy baryons [1]. Interestingly, it was noted that Roper-like excitations have an excitation energy of roughly 500 MeV from the ground state, even in situations where one of the quarks is replaced by a heavy quark. This flavor-independent nature of Roper-like excitations is very interesting. We showed in Chap. 3 that the extra dimensional degrees of freedom of the Sakai-Sugimoto model are also useful in the description of heavy hadrons [2]. We derived the mass spectra of heavy baryons using the Sakai-Sugimoto model in Sect. 3.4. As a further development, we expect a similar analysis to our Chap. 4.

As a further development of the study of hadron resonances in the Sakai-Sugimoto model, we described in Chap. 4 the study of the dynamical nature of nucleon resonances. One of the significant results of this doctoral thesis is that we have pointed out the importance of the aspect of collective motion of mesons for a comprehensive understanding of nucleon resonance. We have performed calculations of the decay widths of one pion emission and have shown that they reproduce well the experimental data for low lying states. A particularly interesting feature is that the relation between the axial coupling $g_A^{NN^*(1440)}$ and the nucleon axial coupling is independent of the model parameters. Such a relation is a feature not obtained from other models. We then performed a calculation of the electromagnetic transition amplitudes. In particular, our analysis captures well the $A_{1/2}$ feature of the Roper resonance (Fig. 4.3), suggesting the importance of the collective motion aspect of nucleon resonances in the description of Roper resonances. On the other hand, the transition amplitudes of the Delta resonance $\Delta(1232)$ and the negative parity resonance $N^*(1535)$ underestimate the experimental data. This may be due to the fact that our calculations are based on the $1/\lambda$ expansion. Therefore, analysis beyond the $1/\lambda$ expansion is expected. Specific analysis is an important issue to be addressed in the future. One possible analysis method is to use asymptotic solutions for instanton solutions as described in Sect. 4.1.2. This produces a contribution from $\tilde{j}^{\mu}_{L/R}$ in the current's expression (4.1.109). Since this contribution is practically the same as the current contribution used in Ref. [3], a large change in the analytical results is expected.

Finally, we remark on several points regarding further future developments.

First, as mentioned immediately above, we expect to improve our results by concretely performing an analysis beyond the $1/\lambda$ expansion. In particular, as we have already mentioned many times, this model, when reduced to 4 dimensions, leads to an action consisting of mesons and baryons, and furthermore, their coupling constants are not parameters but fully determined by the warp factor and the eigenfunctions of the mode expansion, reflecting the QCD information. By clarifying the relationship between the magnitude of these couplings and the asymptotic solution of the instanton solution, the phenomenological aspect of the Sakai-Sugimoto model becomes more clear. As a result of such analysis, a deeper understanding of the dynamic nature of holographic QCD can be achieved.

It is also possible that some of the failures in our analysis arise from problems associated with the definition of chiral currents in the Sakai-Sugimoto model. While the chiral current obtained from the GKP-Witten method, which is a proper tool for holographic QCD analysis, gives a good prediction for the decay width of the one pion emission, it does not reproduce the experimental data of $A_{1/2}$ for the Roper resonance, which exhibits characteristic behavior. Therefore, we have analyzed $A_{1/2}$ of the Roper resonance by considering the 5d bulk theory of the Sakai-Sugimoto model as a 5d hadron effective model like Ref. [4] and by defining the chiral current as a Noether current, and found that we succeeded in reproducing the approximate behavior of $A_{1/2}$ of the Roper resonance. The relationship between these two currents is a theoretical issue to be clarified in the future. For example, in Ref. [5], it is confirmed analytically that the definition of the energy momentum tensor in terms of the Noether current of the bulk theory of the Sakai-Sugimoto model and the definition obtained from the bulk/boundary correspondence exactly coincide in the zero momentum transfer case.

Furthermore, the dynamical properties of resonance states of heavy baryons can be investigated by using our proposed method of introducing heavy flavor into the Sakai-Sugimoto model, as in Sect. 4.3. In particular, Roper-like excitations in heavy baryons exhibit flavor independent properties, thus understanding how Roper resonances corresponding to monopole-like excitations are generated in situations where one quark is replaced by a c quark or b quark is of great interest for understanding the dynamics of low-energy QCD.

Such monopole excitations, which have the same spin parity as the ground state and cannot be captured by a simple picture, are also known in the nuclear physics context, such as the Hoyle state, the first excited state of Carbon, and the $J^P = 0^+$ excited state of Oxygen. The existence of such monopole excitations can be interpreted as suggesting that components strongly bound by a strong force may benefit energetically from monopole-like excitations rather than from excitation of them alone. Such nuclear excitations by solitons have been studied [6–9], and our study may be able to be extended in this direction. It is hoped that understanding these unique excitations to many-body systems governed by strong forces will open up a deeper understanding of QCD.

Finally, there has been great interest in recent years in the study of the gravitational form factor of hadrons, that our research is expected to develop in this direction as well. The gravitational form factor is a form factor derived from the matrix elements of the energy-momentum tensor, and is responsible for information on the mass and spin distribution in hadrons, as well as the distribution of pressure and stress that confine quarks and gluons into hadrons. These quantities, in addition to the electromagnetic and axial form factors obtained in this doctoral thesis, have the possibility to lead to a more detailed understanding of the structure of hadrons. Recently, in Refs. [10, 11], experimental pressure and stress distributions for nucleons have stimulated the study of gravitational form factors, which have been seen only as theoretical objects, and many theoretical studies have been conducted [12–25] (Review : [26]). Furthermore, using the Sakai-Sugimoto model, the expectation value of the energy-momentum tensor of the nucleon has been calculated to obtain the D-term, which is one of the gravitational form factors [5]. One of our future research directions is to analyze the gravitational form factor of nucleon resonances. By studying the gravitational form factor of nucleon resonances, we can understand their internal pressure and stress distribution, which may open a new insight into how the strong force confines the quarks and gluons.

References

1. Arifi AJ, Nagahiro H, Hosaka A, Tanida K (2020) Roper-like resonances with various flavor contents and their two-pion emission decays. Phys Rev D 101(11):111502
2. Fujii Daisuke, Hosaka Atsushi (2020) Heavy baryons in holographic QCD with higher dimensional degrees of freedom. Phys Rev D 101(12):126008
3. Hashimoto Koji, Sakai Tadakatsu, Sugimoto Shigeki (2008) Holographic Baryons: static properties and form factors from gauge/string duality. Prog Theor Phys 120:1093–1137
4. Son DT, Stephanov MA (2004) QCD and dimensional deconstruction. Phys Rev D 69:065020
5. Fujita M, Hatta Y, Sugimoto S, Ueda T (2022) Nucleon D-term in holographic quantum chromodynamics. PTEP 2022(9):093B06
6. Manton NS (2022) Skyrmions—a theory of nuclei. World Scientific, vol 3
7. Halcrow CJ, King C, Manton NS (2019) Oxygen-16 spectrum from tetrahedral vibrations and their rotational excitations. Int J Mod.Phys E 28(04):1950026
8. Halcrow CJ, Manton NS (2020) Tetrahedral intrinsic structure of Oxygen-16 revisited. J Phys Conf Ser 1643(1):012136
9. Lau PHC, Manton NS (2014) States of Carbon-12 in the Skyrme model. Phys Rev Lett 113(23):232503
10. Burkert VD, Elouadrhiri L, Girod FX (2018) The pressure distribution inside the proton. Nature 557(7705):396–399
11. Burkert VD, Elouadrhiri L, Girod FX (2021) Determination of shear forces inside the proton, vol 4
12. Neubelt Matt J, Sampino Andrew, Hudson Jonathan, Tezgin Kemal, Schweitzer Peter (2020) Energy momentum tensor and the D-term in the bag model. Phys Rev D 101(3):034013
13. Chakrabarti Dipankar, Mondal Chandan, Mukherjee Asmita, Nair Sreeraj, Zhao Xingbo (2020) Gravitational form factors and mechanical properties of proton in a light-front quark-diquark model. Phys Rev D 102:113011
14. Choudhary Poonam, Gurjar Bheemsehan, Chakrabarti Dipankar, Mukherjee Asmita (2022) Gravitational form factors and mechanical properties of the proton: connections between distributions in 2D and 3D. Phys Rev D 106(7):076004
15. Kumano S, Song Q-T, Teryaev OV (2018) Hadron tomography by generalized distribution amplitudes in pion-pair production process $\gamma^*\gamma \to \pi^0\pi^0$ and gravitational form factors for pion. Phys Rev D 97(1):014020 (2018)
16. Kim Hyun-Chul, Schweitzer Peter, Yakhshiev Ulugbek (2012) Energy-momentum tensor form factors of the nucleon in nuclear matter. Phys Lett B 718:625–631
17. Kim June-Young, Yakhshiev Ulugbek, Kim Hyun-Chul (2022) Medium modification of the nucleon mechanical properties: abel tomography case. Eur Phys J C 82(8):719
18. Jung Ju-Hyun, Yakhshiev Ulugbek, Kim Hyun-Chul, Schweitzer Peter (2014) In-medium modified energy-momentum tensor form factors of the nucleon within the framework of a π-ρ-ω soliton model. Phys Rev D 89(11):114021
19. Yakhshiev Ulugbek, Kim Hyun-Chul, Schweitzer Peter (2013) Energy-Momentum tensor form factors of the nucleon in nuclear matter in the Chiral Soliton Model. Few Body Syst 54(7–10):1083–1086
20. Mamo Kiminad A, Zahed Ismail (2020) Diffractive photoproduction of J/ψ and Υ using holographic QCD: gravitational form factors and GPD of gluons in the proton. Phys Rev D 101(8):086003
21. Mamo Kiminad A, Zahed Ismail (2021) Nucleon mass radii and distribution: holographic QCD, Lattice QCD and GlueX data. Phys Rev D 103(9):094010
22. Mamo Kiminad A, Zahed Ismail (2022) J/ψ near threshold in holographic QCD: A and D gravitational form factors. Phys Rev D 106(8):086004
23. Polyakov MV, Schweitzer P (2019) Mechanical properties of particles. PoS, SPIN2018:066
24. Lorcé Cédric, Moutarde Hervé, Trawiński Arkadiusz P (2019) Revisiting the mechanical properties of the nucleon. Eur Phys J C 79(1):89

25. Kim June-Young, Kim Hyun-Chul, Polyakov Maxim V, Son Hyeon-Dong (2021) Strong force fields and stabilities of the nucleon and singly heavy baryon Σ_c. Phys Rev D 103(1):014015
26. Polyakov Maxim V, Schweitzer Peter (2018) Forces inside hadrons: pressure, surface tension, mechanical radius, and all that. Int J Mod Phys A 33(26):1830025

SPRINGER NATURE

GPSR Compliance

The European Union's (EU) General Product Safety Regulation (GPSR) is a set of rules that requires consumer products to be safe and our obligations to ensure this.

If you have any concerns about our products, you can contact us on ProductSafety@springernature.com

In case Publisher is established outside the EU, the EU authorized representative is:

Springer Nature Customer Service Center GmbH
Europaplatz 3
69115 Heidelberg, Germany

The manufacturer's authorised representative in the EU is Springer Nature Customer Service Centre GmbH, Europaplatz 3, 69115 Heidelberg, Germany. If you have any concerns regarding our products, please contact ProductSafety@springernature.com

Printed and bound by CPI Group (UK) Ltd, Croydon, CR0 4YY
30/09/2026
02225297-0006